Power Systems

Electrical power has been the technological foundation of industrial societies for many years. Although the systems designed to provide and apply electrical energy have reached a high degree of maturity, unforeseen problems are constantly encountered, necessitating the design of more efficient and reliable systems based on novel technologies. The book series Power Systems is aimed at providing detailed, accurate and sound technical information about these new developments in electrical power engineering. It includes topics on power generation, storage and transmission as well as electrical machines. The monographs and advanced textbooks in this series address researchers, lecturers, industrial engineers and senior students in electrical engineering.

Power Systems is indexed in Scopus

More information about this series at http://www.springer.com/series/4622

Chuzo Ninagawa

Virtual Power Plant System Integration Technology

Springer

Chuzo Ninagawa
Gifu, Japan

ISSN 1612-1287 ISSN 1860-4676 (electronic)
Power Systems
ISBN 978-981-16-6147-1 ISBN 978-981-16-6148-8 (eBook)
https://doi.org/10.1007/978-981-16-6148-8

This Springer imprint is published by the registered company Springer Nature Singapore Pte Ltd.
The registered company address is: 152 Beach Road, #21-01/04 Gateway East, Singapore 189721, Singapore

Preface

Until now, it has become conventional that consumers of commercial power systems can consume as much power as they need. The reason is that the power station issued the power supply command to maintain the grid frequency so that the power company balances the power supply and demand every moment.

However, unstable power generation such as solar power generation has come to occupy some percentage. In the future, it will be difficult to balance the instantaneous supply and demand with the thermal power plant alone and cooperative control with the consumer side will be useful. Since the demand suppression of the power system is equivalent to the increase of power generation, the concept of demand suppression is called as negawatt, and the service system that adjusts the supply and demand including it is called the virtual power plant.

Under such circumstances, the word "virtual power plant" precedes, but if you try to design and develop a service system, it is difficult to catch a book that explains the basic technology in a comprehensive manner. This book is an attempt to explain in detail the basic technology, system design, and performance prediction for practical engineers who actually want to design and develop the service systems. It will be useful for people in various fields such as new IT service start-up, new electric power company management and building facility planning to give an overview of virtual power plant technology.

In this book, we focused on building air-conditioning demand response and power storage batteries as the subject of the resources that make up the virtual power plant. The first reason is that building multi-air-conditioners have 1.5 million units installed nationwide, and if large-scale integrated control is possible, they can become a resource of virtual power plant. Second, it is because the combination with a storage battery is considered to be the key to determine the performance of a virtual power plant. Solar power generation and distributed generators are also resources, but it is believed that the demand response and storage battery control can give the high-speed response performance.

The structure of this book is as follows. The first part outlines Chaps. 1–4 and introduces the general concepts of virtual power plant systems. The construction techniques in second part are detailed in Chaps. 5–8, respectively, detailing the elemental technologies and performances, respectively, and Chap. 9 gives the future prospects that the author considers.

Chapter 1 presents the concept of a virtual power plant against the backdrop of the near future power system issues. Explain what services the system provides. This book describes the reasons why building multi-air-conditioning and storage batteries for power are the main resources of a virtual power plant.

Chapter 2 briefly explains the principle of instantaneous supply and demand balance of commercial electric power system, showing that controlling the increase and decrease of the demand is equivalent to controlling the amount of power generation on the supply and demand balance.

Chapter 3 presents the concept of a computer network systems called a virtual power plant and introduces the defacto standards OpenADR, IEC 61850 and IEEE 1888 standards for smart grid communications specialized for the system development.

Chapter 4 is about fast automated demand response (FastADR) that generates negawatts by controlling the storage battery system for power and controlling the power consumption of building multi-air-conditioning in the office building group with high precision for the elements that make up the virtual power plant and the baseline estimation method that is the basis of weighing negawatts.

Chapters 5–8 describe in detail the elemental technologies needed to realize a virtual power plant system from the viewpoints of researchers and engineers.

Chapter 5 described the charge and discharge control method when incorporating a power storage battery into a virtual power plant. The control model of the power storage battery and the control service communication procedure are described. We also added the storage battery control data transmission delay by mobile data communication.

Chapter 6 describes how to build models of air-conditioning power and room temperature change forecast model by deep learning when high-speed demand response is of a building multi-air-conditioner group. We also show that although the demand response of each building's multi-air-conditioner is uncertain, by aggregating in large amounts, the certainty is improved by the averaging effect.

Chapter 7 describes performance evaluation methods such as responsiveness of the entire virtual power plant system. Specifically, we show mathematical modeling and simulations of demand response performance of multi-air-conditioning demand power and control system.

Chapter 8 explains in detail the implementation technology of communication software for virtual power plant system. Taking OpenADR as communication on the grid side and IEC 61850 and IEEE 1888 as communication on the installation side, the system architecture will be explained at the programming level, and guidance will be given to fill in the gap between the vague function requirements and the formal communication standard.

Chapter 9 summarizes the author's thoughts on the potential resources of the building multi-air-conditioning system, the prospect of system development, and the long-term merits of the virtual power plant system.

Since, author is a member of the "Electric Power Operator-Customer Service Interface Technical Committee for Smart Grid of the Institute of Electrical Engineers of Japan" and introduced the JEC-TR-59002 standard for battery control. We express our gratitude here.

Many people cooperated in compiling this book. Assistant Professor Matsukawa Shun, Joint Researcher of Smart Grid Power Control Engineering at Gifu University, who participated in drafting, Researcher Yoshifumi Aoki, Graduate school students Tomohisa Yamada, Suzuki Keita, Otsu Hideyuki, Kinugasa Hitoshi, Nakayama Takuya, and Asif Iqbal thank you very much.

We hope that this book will be of some help in planning and building a virtual power plant service system.

Gifu, Japan
Spring 2019

Chuzo Ninagawa

Contents

Chapter 1
Smart Grid

1.1 What Is Power System?

1.1.1 History of Power System

The history of commercial power system traces back to 1881 with Thomas Edison's electric lighting company in New York; soon after that, his company installed hydroelectric power generation facility effectively in 1882. However, unlike present power systems, the first commercial power system employed direct current (DC) transmission and distribution. Those systems could not stay alive because the voltage could not easily converted to the high voltage with the available technology at that time. Besides voltage conversion constraint, protection and safety issues in distribution networks and consumer facilities were also concerned. Additionally, massive current flow was there due to low-voltage transmission (thereby causing resistive power losses). Hence, distant power transmission was unfeasible due to consequent voltage drop because of significant resistance of longer transmission lines for greater distances.

Later, in contrast with Edison's DC system, Nicola Tesla developed a poly-phase alternating current (AC) system and AC power system was commercialized with the same mechanism as present. In case of the AC system, the voltage can easily be stepped up/stepped down by incorporating transformers. Therefore, transmitting power over hundreds of kilometers became feasible. The system employs electricity generation at generator-friendly voltage levels, steps up with transformers, transmits at higher voltages, and supplies end users easily handled voltages with step-down transformers. Thereby, the fundamental scheme of commercial power grids became popular throughout global horizons.

Since 1900, the long-distance power transmission plus high voltage transmission became feasible, subsequently improving the net power supply. By then, interconnections of autonomous power grids respective demand areas were possible at same voltage, frequency and phase sequence of AC—known as grid connections in AC power network.

C. Ninagawa, *Virtual Power Plant System Integration Technology*, Power Systems,
https://doi.org/10.1007/978-981-16-6148-8_1

It is to be noted that the prime factors in grid interconnection technology are AC synchronous generators. A synchronous generator is a generator that tends to produce the power system's AC frequency precisely. Keeping in view these characteristics, a large number of synchronous generators distributed in a number of power grids can be stabilized by operating at exactly the same rotational speed and angle, so that the entire grid becomes stabilized. Thus, the foundation for large-scale commercial power systems had been laid, like the present ones.

1.1.2 Supply and Demand Balance of Power System

As described earlier, the basis of a commercial power system is the parallel connection plus operation of a large number of AC synchronous generators. It is, therefore, inevitable to maintain frequency of the entire system accurately and precisely; that if not done, causes interference among the generated power by respective generating units with energy deliverance being ineffective.

Furthermore, in a power system, the voltage can be varied (raised and lowered) by employing transformers. Despite this provision, the AC power system cannot be effective unless the frequency is precisely maintained throughout the power network spanning over hundreds and thousands of kilometers. So, the strict compliance with the system frequency is the core of any large-scale interconnected AC power grid.

Though the mechanism of maintaining system frequency is described in Chap. 2, yet at a glance, if the total power generation capacity and total power demand of the power system match exactly, the supply and demand kilowatts of the entire system maintain the system frequency at specified value.

As shown in Fig. 1.1, the total power generation and total demand of the entire system can maintain the frequency at specified value by balancing both precisely and continuously. If the total instantaneous power generation is slightly larger than the total instantaneous demand, the frequency rises in seconds, and vice versa. This equilibrium, therefore, has to be monitored 24 h a day and 365 days an year at the control centers of Power Companies.

The demand side, is however least concerned about the above mentioned constraints of the generation or supply end of the power system. Therefore, a mechanism has to be required to cater for a wide range of the demand magnitudes, having a firm control over the system frequency. The commercial power system is assumed to be having large-scale AC synchronous generators with power generation output being accurately monitored.

The fundamental property of the AC synchronous generator is that its rotational speed varies with the difference between its generating power and the electric power load, and consequently, frequency of the generated power changes. However, if the load is increased and the frequency gets lowered within a permissible range, the generator's resilience tends to increase its output and to return its frequency to the synchronous one.

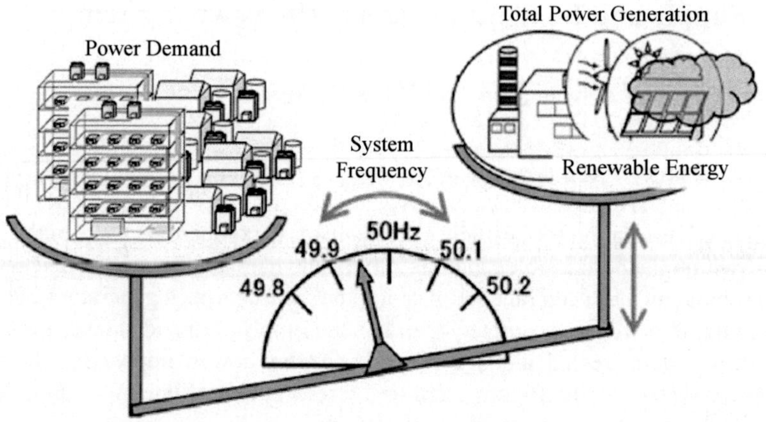

Fig. 1.1 A commercial power system based on an AC synchronous generator maintains grid frequency by instantaneously balancing the total generated power and total power demand of the entire system

As a consequence, when multiple AC synchronous generators are connected to and operated on the same electric power system, there is a property that many synchronous generators can generate synchronized power to match their rotational speeds. This is called synchronization.

Large-scale commercial power system connected at the same frequency, turn into a stable and robust system when tens and hundreds of AC synchronous generators rotate at the same speed. If the scale becomes large, even if there is a local imbalance in supply, it can be absorbed if there are many other generator groups.

Even if some of the generators and/or transmission lines fail and get disconnected, the dropped generated output is compensated by the synchronized power pool of the remaining generator groups, provided that the power shortfall is within certain limits.

Adding more to it, during recent years, supply side change is also dynamic because of significantly growing shares of renewable energy (like solar and wind energy) in the energy pool. The non-consistent power outputs, from renewable sources owing to weather conditions tends to vary the overall energy supply of the system, This behavior produces another set of constraints on the supply end of the power system that are carefully solved by implementing the tangible control algorithms and mechanisms.

1.2 Supply and Demand Control of the Power System

1.2.1 Central Power Supply Control Center

Despite inherent resilience, the generators have certain limitations beyond the resilience range. However, synchronous generator groups are incorporated with well-coordinated restoration and resilience mechanisms to compensate some of the imbalances (within the permissible range) in supply and demand. In general practice each power company has a central control center that instructs each generator to generate power output in order to accurately maintain the supply–demand balance throughout the power system. Technically, it is termed as central power supply control center in the power system world. Figure 1.2 shows screen output of one of the Japanese (as an example) central power supply control center.

The modern large-scale commercial power system spans over vast geographies. The individual generator outputs have to be monitored plus adjusted time to time in order to address the aforementioned constraints. This is accomplished either by individual generator feedback and control or centrally by central power supply control center.

The earlier employs local control mechanisms whereby central communication turns out to be unfeasible from the central power supply control end. The latter employs the output command from the central power supply control center to be followed up and responded by the units with employing certain protocols. It is to be noted that all control algorithms are tailored to equilibrate generation-consumption extrema, alongside precise frequency control within permissible frequency range.

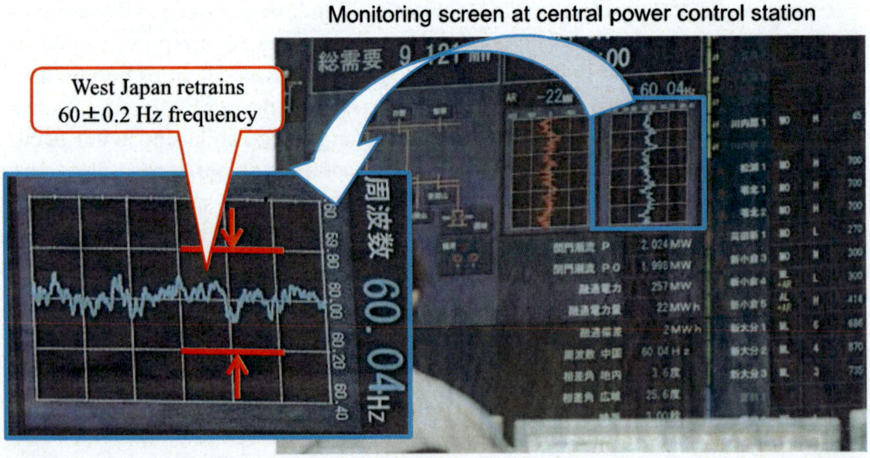

Fig. 1.2 On the monitoring screen of the central power supply control center of the commercial power grid, monitoring the grid frequency and commanding each power station to adjust the power generation output (Modified from the article image of the Japanese Electric Newspaper August 17, 2015)

In any case, so far, as the total power generation and total demand power can be balanced instantly. It has been the paradigm, until now, that the power supply and demand control is performed with power plants alone regardless of how demand fluctuates.

The traditional power plants managed by the central power supply control center include hydropower, oil/LNG/natural gas/coal-fired, and nuclear power plants. In order to equilibrate generation and supply, rapid controls within fractions of second or at maximum within few seconds are required. In order to do that a well-equipped central power supply control center is needed to cater for the abruptly rising and falling in power.

The hydropower stations are well suited for the central control, by central power supply control center, because of their quick response in terms of speed and output. However, the total installed capacity is limited. Furthermore, in case of pumped storage power plants, another constraint lies there in terms of the limited storage capacity with consequent power generation constraint i.e. short lived compensation lasting only for few hours.

Hence, the reliance over oil, natural gas and other thermal power generation groups by the central power supply control center in order to maintain the demand and supply balance is obvious. Keeping in view the fossil fuel consumption, the environmental concerns and fossil fuels shortage have to be addressed and mechanisms ought to be developed for least dependence over these resources in near future.

In recent years, renewable energy power generation, such as hydroelectric power generation and thermal power generation, in which supply and demand balance control cannot be commanded from the central power supply command center, has become widespread. Therefore, the proportion of traditional power plants as described above will decrease relatively. Then, in addition to the decrease in synchronization power, the essential problem of having to control the balance between supply and demand, including unstable output fluctuations, has emerged.

1.2.2 Renewable Energy Generation and Supply and Demand Control

The aforementioned solar wind power generation, renewable energy generation, having momentarily fluctuating inconsistent outputs, are termed as natural energy power generation. Renewable energy generation, over here, generally refers to large scale centrally controllable power plants such as hydel ones.

The traditional power systems (based on the concept of synchronized energy) employ central frequency control by adjusting respective outputs of the large scale synchronous generators. Such parallel operation of generators stabilizes the entire power system. However, it is feasible at significantly lower shares of renewable energy generation in total power generation.

In solar power generation, the daylight is the decisive factor in terms of the power output. This is due to greater intensity of sunlight during the daytime and lower otherwise. The output is, therefore, significant during the daytime particularly during peak sunshine hours while rapidly lost with sunset. Although it contributes to the overall power pool, but the set back is in terms of all-at-once (post sunset) disappearance of the solar power. Thus, with widespread solar power generation in commercial power grids, the share of renewable energy generation in the overall power generation drastically fluctuates depending upon the time zone even during the same day.

Figure 1.3, known as "Duck Curve", depicts the problem of solar power generation and balance adjustment, analyzed by the U.S.A. system operation agency, CAISO (California System Independent Operator) [1]. The horizontal axis is the time/hour axis with vertical one depicting the power generation capacity of the thermal power plants to balance the generation and consumption. From 2012 to 2020, drastic changes in thermal power requirements during the day versus evening hours is noticeable.

Among renewable energy, solar energy generation contributes appreciable share to the overall power needs during daytime. During peak sunshine hours, when solar radiation is intensive whereas energy demand is relatively low. The possibility of solar power alone exceeding the total power demand is important.

In such case, the thermal power generator groups will have to be suppressed on a large scale. Alongside this limitation, another constraint of rapid fall of solar energy around sunset hours is eminent. To maintaining the supply–demand balance, the thermal power generator groups have to be turned on immediately. Consequently, the thermal power plants need to be turned on and off at comparatively rapid rates.

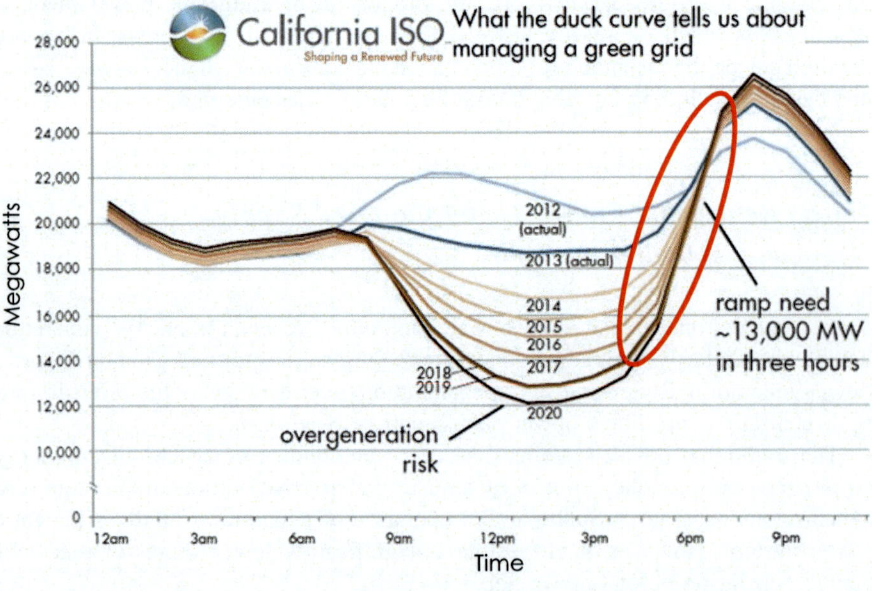

Fig. 1.3 Duck curve where California Operations Network CAISO warns, adopted from [1]

The duck curve for thermal power generation is changing year by year, as in 2012 the requirement for thermal power generation from early morning to daytime is recognizable. However, with increasing solar power generation, the solar power output becomes sufficiently large to reduce the energy needs during daytime.

Hence, during the daytime, thermal power generators output essential to be cut short to an extent, where, there is a risk for entire demand to be met by the photovoltaic plants alone i.e. during the peak sunshine hours with minimal energy requirements. This is termed as over generation risk as shown by the duck curve diagram in Fig. 1.3.

It may seem desirable to meet the power needs of the whole system with environment friendly natural energy. However, as mentioned earlier, in the power system, the instantaneous consumption-generation balance is an absolute requirement whereas in renewable energy power generation, variations in meteorological conditions directly cause the fluctuations in electric power output. Therefore, it is impossible to control solar power generation to meet momentary changes in power demand, which CAISO warns of as “risk”.

Similar is the case in Japan, because of the massive photovoltaic installations in 2019. Such conditions generally occur during spring and summer holidays when the solar illuminance is good while the power demand is relatively small. Most drastic, the case of Kyushu Electric Power (a Japanese local power company), where the solar power generation of 6.21 GW at 1:00 pm on May 3rd 2018 contributing 81% of the total power generation within the Kyushu Electric Power area according to [2]. The ultimate duck curve eventually hits Japanese commercial power system, that is shocking for the power system experts.

The thermal power generators stand by until the energy demand is met by renewable energy generation. But since sunset, the output acceleration of thermal power plants (ramp up) is inevitable. However, two major problems are faced with this; firstly, the economic one due to the decrease in thermal power generation (owing to lesser operation) and secondly the rapid output acceleration required for the ramp up operation.

In past, when power generation by renewable energy such as solar power generation was not so much, the problem of thermal power generators with a short operating time was pointed out. For example, as shown in Fig. 1.4, the rated capacity of a thermal power plant operating less than 80 h a year, i.e. less than 1% of the total yearly time, the total value was 3.84 GW as in [3]. In other words, since the rated capacity of a typical nuclear power plant is 1 GW, it means that it is necessary to maintain the thermal power plant for four nuclear power plants even if they operate only for a few days a year.

No matter how much renewable energy generation share in the power pool, the inconsistent output owing to meteorological conditions urges the need of thermal and hydroelectric power plants in order to balance the consumption and generation.

The objective of all of the above discussion is based on that the instantaneous supply–demand adjustment of the power system is mainly deal with thermal power generators. Here, with respect to this major reason, the author would like to consider the point of view that the balance between supply and demand has the same effect by adjusting the demand power. If the demand power of the grid can be adjusted

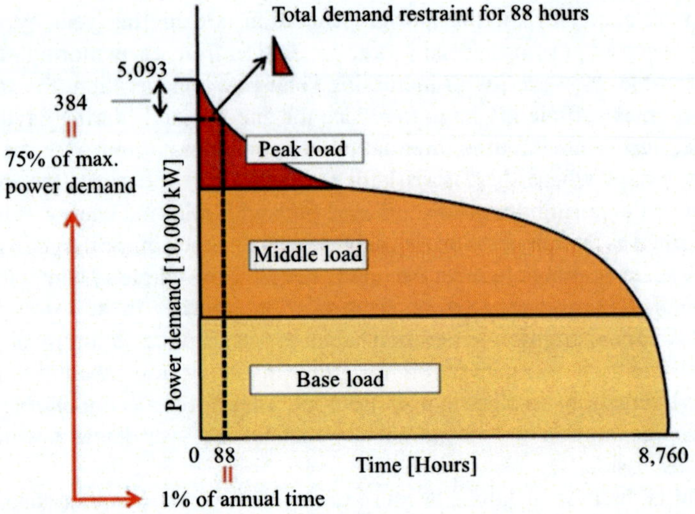

Fig. 1.4 Some thermal power generators operate only for limited time due to grid supply–demand adjustment, adopted from [3]

with the same accuracy and speed as the adjustment by thermal power generators, it should be equivalent to the adjustment of generated power. In other words, it may not be necessary to balance the instantaneous supply and demand simply by adjusting the power generation output of the thermal power generators. In future, there may be a scheme that balances the instantaneous supply and demand by implementing the control on the power demand side, as far as, economy and reliability enhancement.

Despite predominant merging, the definition and concept of the smart grid is somewhat diverse, such as the sophisticated utilization of IT and advanced control technology in power systems. However, in author's opinion, the essence lies in the paradigm shift from adjusting large-scale concentrated power plants to employing large dispersed power demand control.

1.3 Smart Grid

1.3.1 Fast Demand Response

It's not true that, so for, there has not been such techniques to manage and adjust the power demand for the power system. For example, for each time zone such as late-night electricity charges, there is a time-leveling measurement scheme of power demand by electricity unit price. However, virtual power plants can be better understood if they are considered separately from the demand response in the smart grid described in this section.

Table 1.1 Traditional demand control and smart grid ADR

Item name	Traditional demand control	Smart grid ADR
Automation	Control inside customer equipment	Service provider controls communication
Time granularity	30 min power unit	Minutes possible
Control target value	Fixed contract	Real time
Control trigger	Set in advance in comparison with contract power	Real-time communication from server
International standard	None	OpenADR etc
Aggregation (multi-building management)	Rare	With cloud aggregation

In addition, demand control refers to suppress the demand so as not to exceed the upper limit. The upper limit of demand is monitored and suppression for each 30 min time frame. It is not automatic control that makes it follow the target energy every moment like feedback control.

The current demand control is a static method in which each customer not to exceed from assigned power. ADR (automated demand response), is also in the experimental stage which dynamically responds to real-time power rates and emergency power reduction; assuming future smart grids. Future building energy management systems (BEMS) may be required to respond to such ADRs, see Table 1.1.

In this section, we will consider ADR, which automatically controls demand in response to the high speed of the smart grid era. In addition, we will look into future systems that enable ADR aggregation to communicate with a large number of customers remotely.

It is alleged that suppressing power consumption by automatic demand response is equivalent to power supply unit in terms of power balance. So that, there are three types of power supply services.

The first is energy service that adjusts the actual energy balance. The second is capacity service that is on standby with the capacity available when needed. The third is ancillary service that maintains power quality such as frequency.

Demand response can be classified into manual demand response, automatic demand response and fast automatic demand response (hereafter, FastADR). At present, they are rarely distinguished, but they have the following characteristics.

Manual demand response is slow and time-consuming due to human interference. So, it seems that it is not suitable for aggregation that aggregates a large number of customers remotely. Even if aggregation is performed, it can manage with fixed contract peak cuts and the power market on the next day, but it cannot handle with services that require high-speed response such as ancillary services.

Next, because automatic demand response is based on machine to machine (M2M) communication between computers. It is suitable for ADR aggregation where administrators are unnecessary and a large number of consumers are remotely aggregated. In ADR, since request, compliance, and load control are automated by computer, it

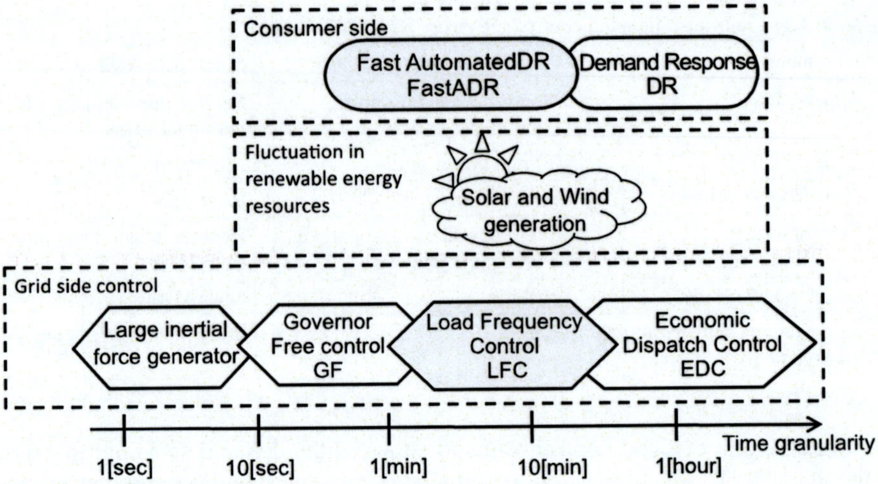

Fig. 1.5 Allocation by time granularity of grid frequency control of electric power system and correspondence of time granularity of natural energy generation fluctuation and suppression of high-speed demand response demand

can cope with frequently occurring service. In addition, it is possible to aggregate the negative watt (hereafter, negawatt) production of consumers directly dispersed widely between computers. To be used in a virtual power plant system, it must evolve to at least the level of ADR.

Furthermore, FastADR is an automatic demand response that responds quickly in minutes or seconds, and is still in the research and development stage. Figure 1.5 shows the time granularity of different methods of gird control. As a virtual power plant, large-scale aggregation of FastADR is essential in order to achieve system control participation by receiving an output adjustment command from a central power supply command station as in the case of a conventional thermal power plant.

1.3.2 Virtual Power Plant System

At present, generally considered "virtual power plant", the resources that constitute it are conventional generators, but we intend to achieve them by merging distributed units on a large scale. As of 2019, various virtual power generation demonstration projects seem to indicate the meaning of merging the state-of-art distributed generators.

In this book, the new virtual power plant not present aggregation of distributed generators but a new framework. It is a virtual power plant of a new framework by merging a group of building multi-type air-conditioning systems and the group of auxiliary storage batteries, and it can be called FastADR virtual power generation system.

Distributed generators that use fuel, such as small diesel generators, are understood as emergency power sources, but normal operation may be a problem from a long-term perspective such as global warming and fossil resource depletion.

In addition, if it is a large-scale heat storage system etc., with above-mentioned facility alone, the up/down load power source by the heat storage tank will be able to hold the power control negawatts with high accuracy for a long time. Large-scale equipment has been introduced as a regional heating and cooling supply system for several decades. On the other hand, FastADR, which is power consumption control based on power suppression command from the grid side, is a supply and demand adjustment as a future income source to replace the peak adjustment contract with the electric power company and will be an important element of the virtual power plant. However, the virtual power plants in this book that aim for high-speed response are excluded.

In addition, in principle, the battery power supply can be made responsive promptly as well and energy can be stored, so it can respond to power suppression command. In terms of performance, it can be an ideal virtual power plant component. However, as of 2019, the market value for the high-speed response of the virtual power plant has not been seen financially, so large-scale capital investment is a problem. However, if there is a limited capacity, the investment amount will be within the possible range, so it will be possible to invest in the real-time supply and demand adjustment market if it becomes clear the kW value that responds quickly.

Solar power and wind power are good prospects of energy sources for virtual power plants, but it is difficult to control due to variation in weather conditions. Of course, it has been found that if a large amount of distributed solar and wind power sources are aggregated, the short-term uncertain fluctuations are mitigated by averaging effect.

As a component of a virtual power plant, it is a method to immediately suppress the demand for power by FastADR, which has not been studied so far. The power source such as a small generators are originally similar to the scheme in which the commercial power system manages and controls as the power generation side, so the timing to start up differs from the conventional one.

In the case of a virtual power plant as shown in Fig. 1.6, the author thinks that the core of the paradigm shift is to engage in the demand and supply control of the system by precise high-speed control of the power demand, which was 100% passive in the past. Therefore, we describe the FastADR aggregation targeting load, which has not been a major player in the past as the core of the virtual power plant system.

There is a problem of load response characteristics of individual customers, even if it is attempted to aggregate and control a large amount of aggregation just by suppressing the power consumption of the target load. Therefore, there is a concept to compensate for FastADR's response speed delay and control residual, which remain large even if aggregated in a large amount, and FastADR response uncertainty due to customer's convenience by combining with a minimal battery system.

In this book, in order to highlight the core of this paradigm shift, we focus only on customer load and battery system. A large-scale FastADR and storage battery system will be integrated to aim at a negative power generation system.

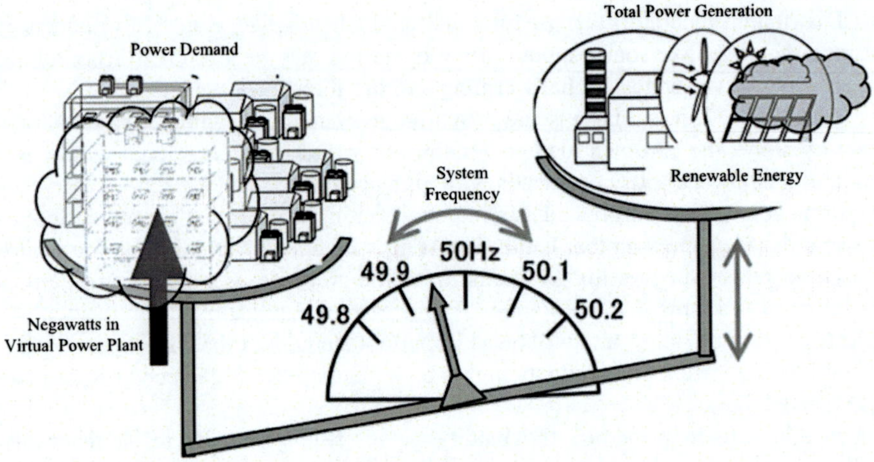

Fig. 1.6 A virtual power plant can reduce the demand for electricity and generate negawatts, and the balance between supply and demand can be obtained equivalent to the increase in power generation

1.3.3 Air-Conditioning Power Demand Response

When considering the target load of demand response, demand response controllability and sense of scale seem to be demand factors.

In this book, we consider that the power consumption is promptly responded in minutes to the demand for power consumption increase/decrease from the grid which is the target load of FastADR as a resource of the virtual power plant.

Generally, in order to respond to the demand at the contract meter interface of the customer unit, there is a broad demand response in which power is generated by the distributed power supply in the customer and the power consumption of the entire customer is reduced accordingly. However, in order to make it easy to understand, this section describes the target load for demand response in a narrow sense, in which the power consumption of the power load is purely controlled promptly and the power consumption is temporarily reduced.

There is no need for lighting load, elevator load or outlet load for pure load equipment that can respond promptly as demand response according to the demand of customer load system. However, they are often noticed by customers immediately, and it is said that demand response controllability is insufficient. On the other hand, air-conditioning equipment is intended to change relatively slowly, that is, the temperature of the indoor space. There is a heat storage effect due to the heat capacity of building structures etc. Therefore, if the duration is short and the range is small, there may be a range of power suppression where the comfort hardly deteriorates. In this book, we simply call it "demand response controllability".

An example of air-conditioning equipment is illustrated in Figs. 1.7 and 1.8 shows the configuration of multiple air-conditioners into a typical office building.

Fig. 1.7 Installation example of building multi-type air-conditioner

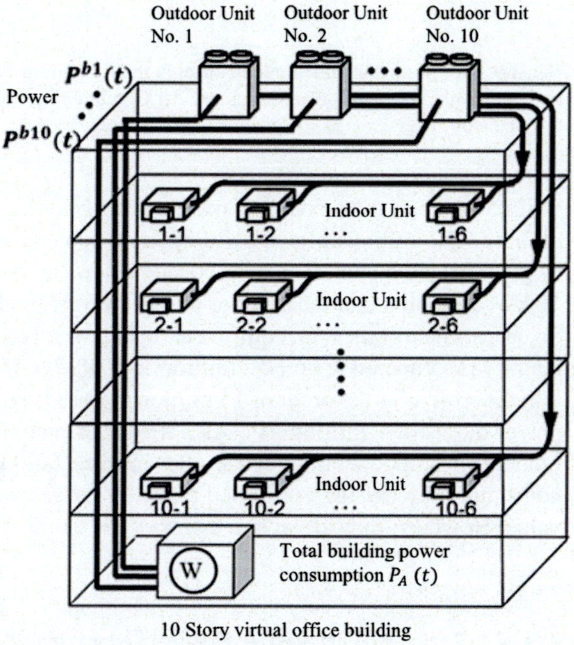

Fig. 1.8 Configuration of multiple air-conditioners installed in a typical office building

The demand response by air-conditioning may of course also be considered for home use, but the power load scale of each unit is too small. Therefore, the response accuracy to demand response seems to be low. Therefore, in this document, we will deal with only large-scale air-conditioning equipment for business use. This book focuses on building multi-type air-conditioning among the commercial air-conditioners. Building multi-type air-conditioners are the type of decentralized air-conditioning facility. These types of air-conditioners are widely installed in Japanese commercial buildings. In this book, author considered the Japanese equipment as an example. Regarding the penetration rate, the number of domestic (in Japan) shipments of building multi-type air-conditioners increased in this decay, as shown in Fig. 1.9. At present, annual shipments to Japan are over 100,000 every year.

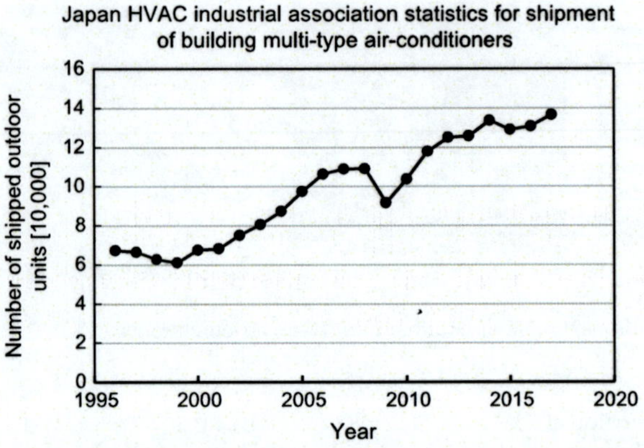

Fig. 1.9 Domestic annual shipments of only building multi-type air-conditioners among package air conditioners (Japan Refrigeration and Air Conditioning Industry Association official statistics: Created from www.jraia.or.jp/cgi-bin/stat/detail.cgi?ca=1&ca2=3)

Taking Japan as an example, we estimate the order of the total power consumption of these 1.52 million units of building multi-type air-conditioners in Japan according to [4]. The rated cooling power consumption can be assumed to be approximately 10 kW, assuming that the outdoor unit is uniformly rated cooling capacity of 28 kW. So, an estimate of the total equipment rated power consumption installed in the whole country is estimated to be 1.52 million $\times$ 10 kW $\approx$ 1500 $\times$ 10,000 kW. It can be said that the scale is of the order of 15 million kW, 15 GW. This corresponds to 15 nuclear power plants of 1 million kW. As a matter of course, of the above, the ratio of the number of units operating at the relevant time, and the load factor for rated power consumption make the consumed power at the relevant time a fraction of the above value. However, they are aware that they are candidates for huge demand response.

1.3.4 Air-Conditioning Power Demand Response and Storage Battery

The building multi-type air-conditioning system is mainly based on the "down" demand response that suppresses the power demand. The "raised" demand response is exceptional as long as it is not a storage accessory. In addition, even if the degree of importance is different, the air-conditioning cannot be operated after at least some need arises. So, the limitation of the target air-conditioning, the suppression range, and the duration time are limited. In addition, "Rebound" may occur, if the control value width duration is large, the air-conditioner accelerates the air-conditioning capacity to return to the original comfortable room temperature at the end of FastADR, and the power consumption is consumed more than the demand

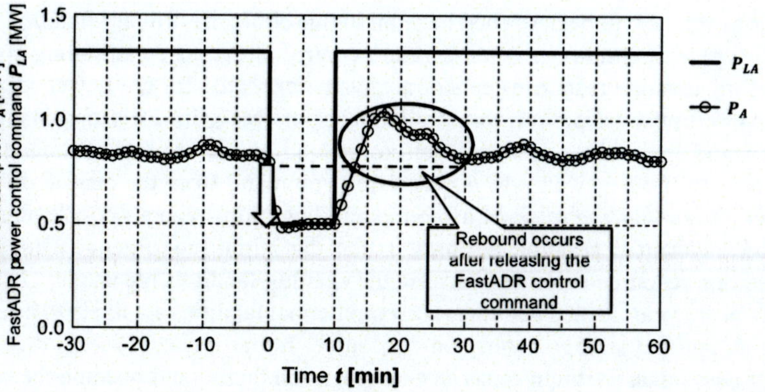

Fig. 1.10 Simulation example of building multi-type air-conditioner FastADR response delay and termination rebound when power storage battery system compensation control is applied, adopted from [5]

response start time. An example is shown in Fig. 1.10. This example is a simulation example as in [5], in which FastADR is superimposed synchronously on 100 air-conditioners, and shows the time change of the total power.

Of course, there is a method of defining demand response group and suppressing power by rotation number, distributing release of demand response appropriately, or gradually returning power suppression release to ramp shape. They will be described in detail in the development techniques section later in this book.

However, from a global point of view, when a large-scale virtual power plant is configured from FastADR aggregation of building multi-type air-conditioning, in order to ensure that it can cope with the increase command and to secure the follow-up performance to the output command signal.

The quality is managed by controlling the response speed, duration, increase/decrease accuracy, and post-end rebound. In this way, by controlling the power control negative wattage to be executed exactly as required, it distinguishes it from the conventional demand management technology. The simpler "simple power saving" is better just by lowering the power anyway, and there is no idea of maintaining the reduced kW precisely. In addition, the conventional 30 min power demand "demand control", which is the next step, is to control only the upper limit for maintaining power. Also, the recent term "demand response: DR" is used including slow power demand control that reserves from the previous day in hourly units.

On the other hand, the negative power generation method called FastADR is different from those conventional demand management control. It is controlled by direct communication between computers, and it is high speed and automatic such that control accuracy is required both up and down to the generated negawatt command level with respect to the target negawatt command level in minutes. It is desirable for a virtual power plant that demand response, that is, technology called FastADR is used for power demand control of a building multi-type air-conditioner.

However, as will be described in detail in this book, the primary function building multi-type air-conditioner is temperature control in the air-conditioning space, and therefore accurate responses cannot always be expected for the power suppression FastADR for that purpose. It would still not be possible to have a return on investment that would make the response level correspond exactly to the command level in response to the negawatt power generation command from the central dispatching center. Originally, the merit of using the building multi-type air-conditioning group as the negawatt generation resource is that the initial investment as the negawatt generation resource is limited because the existing facilities are used.

In recent years, office buildings and commercial facilities are likely to be equipped with an optimal storage battery power source. In particular, it seems that after the Great East Japan Earthquake, as an example, each facility has become the most suitable source of emergency power, changing from the past. Also, in order to protect data that is more valuable to companies than ever, there is an increasing number of cases where storage battery-type emergency power supplies are also used as uninterruptible power supplies. It is a facility that survives a power failure of about 50 kVA for about 10 min or about 30 min.

It is possible to use a part of the stored energy to help the FastADR control error of the building multi-type air-conditioner group. Of course, a large amount of charge and discharge that harm the purpose of the emergency power supply is a problem. Figure 1.11 shows such large power storage battery system for industrial emergency power supply.

However, as will be described later in this document, it is possible to make the building multi-type air-conditioning group follow the negawatt control command level from the power system side by skillfully integrating FastADR with advanced ICT and AI technology.

Furthermore, in order to secure the accuracy and certainty of Negawatt power generation as a virtual power plant captured from the grid side via an aggregator, it is an effective approach to use a combination of power storage battery systems in a limited manner.

Therefore, the key mechanism is to share the above-mentioned quality assurance in any hierarchy of the large-scale pyramidal hierarchical structure of the virtual power plant. In this document, in the following chapters, the construction technology of a virtual power plant that integrates the building multi-type air-conditioning facility group and the power storage battery group and totally controls these distributed resource groups to become equivalent to an integrated generator.

Fig. 1.11 The appearance of a large power storage battery system for industrial emergency power generation (Reprinted from the homepage image of GS Yuasa)

References

1. California System Independent Operator (CAISO). http://www.caiso.com/Documents/Briefing_DuckCurve_CurrentSystemConditions-ISOPresentation-July2015.pdf#search=duck%20curve
2. The Denki Shimbun: Preparing for output control: Electricity supply and demand results during the long holidays in the Kyushu area (1:00 pm), June 6, 2018
3. Agency for Natural Resources and Energy. http://www.enecho.meti.go.jp/about/whitepaper/2015html/3-2-2.html
4. Japan Refrigeration and Air-conditioning Industry Association, *Domestic Shipment Statistics of Air-conditioners*. http://www.jraia.or.jp/cgibin/stat/detail.cgi?ca=1&ca2=3
5. C. Ninagawa, *Dynamic Equilibrium Rebound Suppression Method for Building Multi-type Air-conditioners Fast ADR Aggregation* (Institute of Electrical Engineers of Japan Smart Facility Study Group, SMF-19-016, 2019)

Chapter 2
Power Supply and Demand Balance

2.1 What Is the Supply and Demand Adjustment of the Power System?

2.1.1 Frequency in AC Power System

In previous chapter, we mentioned that the basis of the commercial power system is the accurate operation of the AC synchronous generators. In this chapter, we will explain the relationship between the precise operation of the AC synchronous generators and the instantaneous supply–demand balance.

In power system, all the AC synchronous generators are operating at the same rotational speed, and the system frequency is determined by maintaining the rotational speed. The frequency fluctuates with the balance between the power consumption (demand) of general households, companies, and factories, and the power generation amount (supply) of power generation facilities such as thermal power generation and hydroelectric power generation.

Therefore, if the amount of power generation and the amount of demand are balanced, the frequency will be stable. On the other hand, when the amount of power generation is insufficient relative to the amount of demand, the frequency also decreases slightly. In addition, when the amount of power generation becomes excessive, the frequency slightly increases. The amount of change is usually on the order of 0.1 Hz in the stable state.

Below, in order to explain the transient demand–supply balance and the principle of frequency change, we will briefly look at the basics of power engineering along with textbooks [1]. First, we show in simplified form how the difference between the power output of an AC synchronous generator and the electrical load changes the rotational speed. The power system as a whole is a combination of these one by one movement.

C. Ninagawa, *Virtual Power Plant System Integration Technology*, Power Systems,
https://doi.org/10.1007/978-981-16-6148-8_2

First, consider the movement of one generator. P_m [MW] is a mechanical input applied to the rotor from a prime mover such as steam turbine of a thermal power generator, and the power consumption (load) of the connected power grid, i.e., the electrical output of the generator is P_e [MW]. Assuming that the angular velocity of the root is ω [rad/s], the equation of motion of the generator is expressed by Eq. 2.1.

$$I\omega\frac{d\omega}{dt} = P_m - P_e \tag{2.1}$$

Here, I represents the moment of inertia [kgm^2].

Also, assuming that the rated angular velocity of the rotor is ω_n [rad/s], kinetic energy W [J] when rotating at the rated angular velocity is also rated at the rated angular velocity of the rotor ω_n [rad/s], and rotates at the rated angular velocity. The kinetic energy W [J] is expressed by Eq. 2.2. And we call twice the W the inertia constant M [MWs] as Eq. 2.3.

$$W = \frac{1}{2}I\omega_n^2 \tag{2.2}$$

$$M = I\omega_n^2 \tag{2.3}$$

Equation 2.1 can be expressed by the inertia constant as Eq. 2.4.

$$\frac{M}{\omega_n} \cdot \frac{d\omega}{dt} = P_m - P_e \tag{2.4}$$

According to Eq. 2.4, when the mechanical input P_m is larger than the electrical output P_e, the rotational speed of the generator increases and the frequency also increases. Conversely, if the mechanical input is smaller than the electrical output, the frequency decreases. Also, the larger the inertia constant M, the slower the fluctuation of the rotational speed relative to the fluctuation of the input/output balance. A generator with such characteristics controls the turbine etc. to keep the frequency constant against fluctuating power consumption.

Strictly speaking, each generator can maintain its steady state at a constant frequency for the entire system even if it is rotating at a slightly different speed from the rated synchronous frequency of the system in a very small range. Here, if it is simplified to explain the principle and it is approximated that all the operations of each generator are synchronized. The difference between the sum of the power generation output of the whole power system and the sum of the demand power, i.e., the supply and demand balance of the whole system is represented by Eq. 2.5.

$$\sum_{k=1}^{N}\frac{M_k}{\omega_n}\frac{d\omega}{dt} = \sum_{k=1}^{N}P_{m,k} - \sum_{k=1}^{N}P_{e,k} \tag{2.5}$$

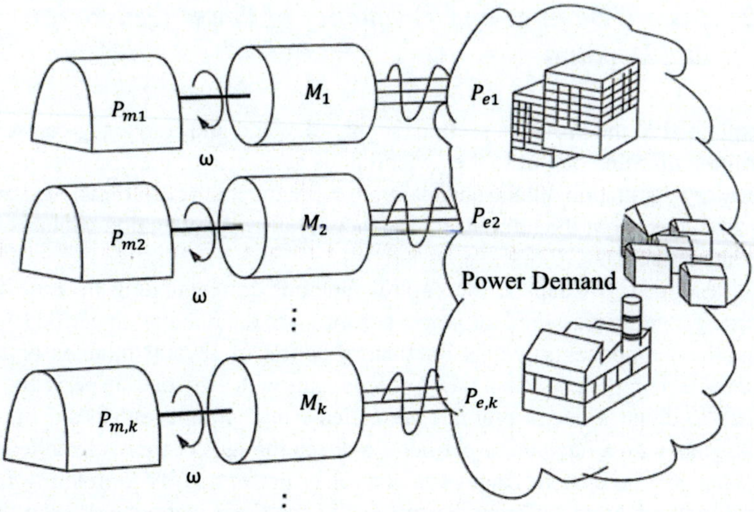

Fig. 2.1 Operation simplified diagram of grid-connected operation of AC synchronous generator description of Eq. 2.5

As shown in Fig. 2.1, the inertial force of many synchronous generators is balanced with the demand–supply balance of the whole system, and the fluctuation of the unified rotational speed of the whole system, that is, the fluctuation of the frequency of the whole system is determined. In other words, the fact that the grid frequency does not change stably means that the common rotation speed ω of the generator group does not change. For simplicity, we ignore the losses due to transmission and distribution, and we treat here that the generator's electrical output is equal to the demand power.

The supply–demand balance state means that the left side of the Eq. 2.5 is zero, and the mechanical output (total power generation amount) and the electric output (total demand) of the entire system on the right side are equal. In other words, it can be seen that the balance between supply and demand of the entire system determines the rate of change of the rotational speed due to the inertial force of the whole generators group.

Thus, whether the frequency of the whole system suppresses the fluctuation as the inertial force against the fluctuation of the balance of supply and demand is determined by the sum of the total inertia constants of the generator group. Therefore, in recent years, when the amount of natural energy generation such as solar power generation increases and the number of synchronous generators such as thermal power generators decreases, the synchronous thermal power of the entire system decreases. And the stability of the entire system decreases.

2.1.2 Instant Balance and Frequency of Power Generation and Demand

As explained in the previous section, supply–demand and frequency control are to control the machine output to the synchronous generator of the power plant against momentary fluctuation of load and maintain the number of revolutions. And maintain the grid frequency at the reference frequency. This load fluctuation contains various periodic components.

As described in Chap. 1, the control methods corresponding to these fluctuations are generally classified into three according to the fluctuation period (Fig. 2.2). Fluctuation components with a fluctuation period of several minutes or less are called cyclic components, fluctuation components with a fluctuation period of several minutes to about a dozen minutes are called fringe components. And fluctuation components with a fluctuation period of a dozen minutes or more are called sustain components. The system frequency control is performed by combining multiple control methods with different control cycles for these components with different fluctuation cycles.

For period components, the fluctuation range is also small, and can be absorbed by GF (Governor Free) operation, which is control by each generator itself. The amount of fluctuation also increases for the fringe component, so it cannot be adjusted with GF operation alone. Therefore, control is performed by LFC (Load Frequency Control) that detects frequency deviation and load fluctuation, adjusts the output of the generator in real time and absorbs the fluctuation as in [2]. The sustain component has a fluctuation period of more than a dozen minutes, and since the amount of power generation is large and it is possible to predict fluctuations in advance. Therefore, EDC (Economic Load Dispatch Control) is used, which can distribute power efficiently based on this prediction and can reduce the cost of power generation.

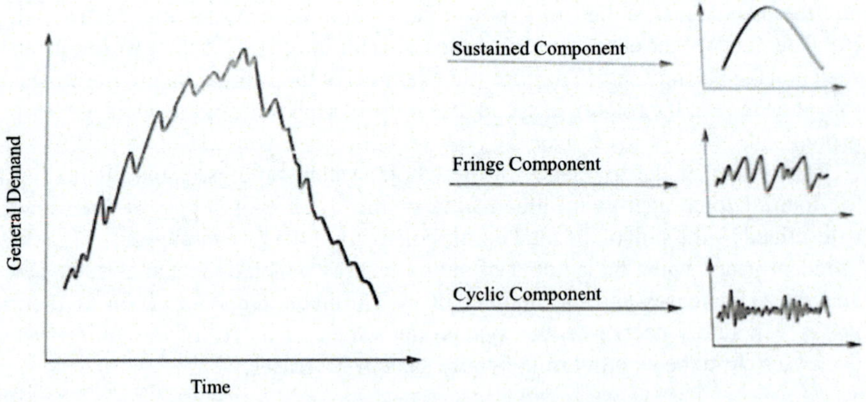

Fig. 2.2 Divide the fluctuation of total system demand by its time component

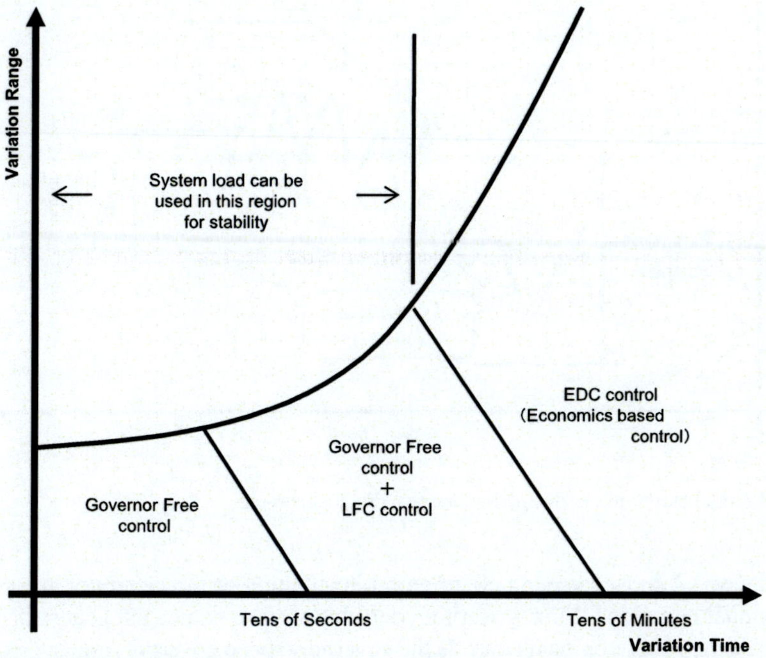

Fig. 2.3 Distribution of control to specified value of grid frequency by fluctuation time component

An overview of these allocations to the time axis of supply and demand balance control is shown in Fig. 2.3. This division is not necessarily clearly divided into independent divisions, but a single generator is mixed with a plurality of control output signals to determine the final operation amount to generator power.

2.2 Role of Central Power Supply Control Center

2.2.1 Load Frequency Control LFC

As outlined in Chap. 1, the central dispatching center within the utility company is a facility that monitors and controls the entire power network. In other words, the central power supply command center constantly monitors the power grid and issues control commands for output adjustment to each generator according to the ever-changing demand–supply balance. And maintains the power-supply balance. Large-scale thermal power plants are also constantly controlled from the central power supply control center to adjust the output.

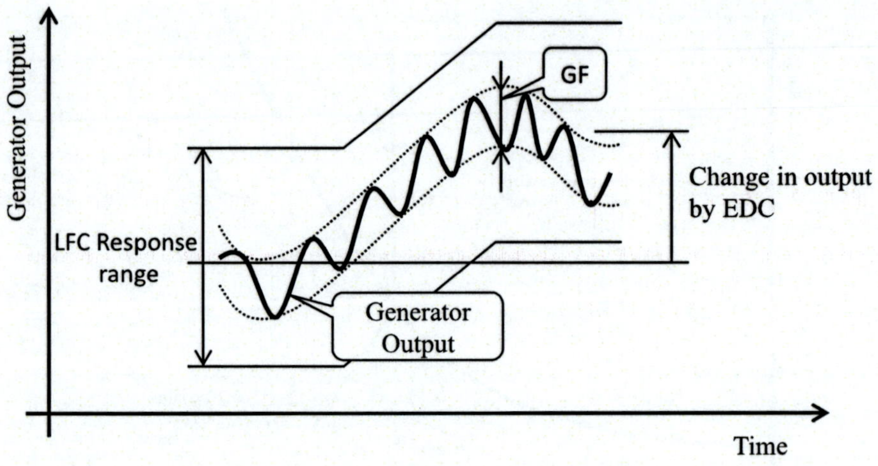

Fig. 2.4 Generator output fluctuation due to individual control

Figure 2.4 shows an example of output fluctuation of the generator according to individual command. First, governor control is performed in each generator. This is because the generator itself controls the governor (speed governor) and the rotational speed of the generator is kept constant. Specifically, in thermal power generators, when the frequency rises, the valves are closed to suppress the amount of steam to the turbine, and when it decreases, the valve is opened to increase the amount of steam.

There is a characteristic as shown in Fig. 2.5 between the power change and the frequency of each generator and load. Here, f_N is the rated frequency (reference frequency) or rated rotational speed, f_0 is the no-load rotational speed (no-load frequency), and P_{G_r} is the rated power of the generator.

As the output of the generator rises in frequency, the governor works and the mechanical input to the generator decreases, causing the electrical output to decrease.

Fig. 2.5 Power frequency characteristics of generator output and load

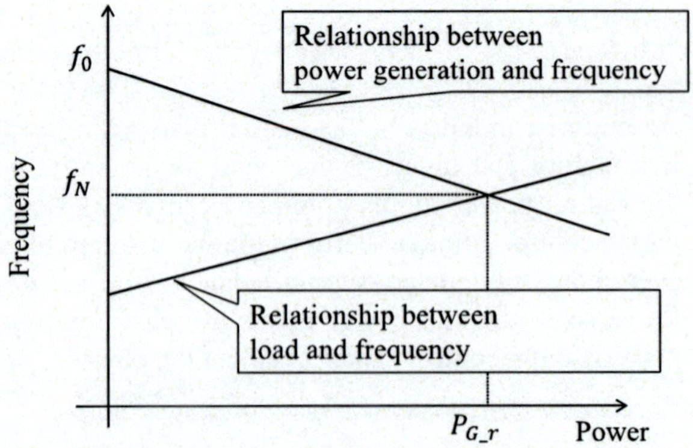

Conversely, when the frequency decreases, the mechanical output increases, and the electrical output also increases.

The speed arbitration rate ε, which indicates how much the generator speed changes when the frequency of the connected system changes, is expressed by Eq. 2.6.

$$\varepsilon = \frac{f_0 - f_N}{f_N} \times 100 \tag{2.6}$$

As ε is smaller, the frequency is more stable, but too small, it is not desirable for a generator because the electric output fluctuates largely due to slight frequency fluctuation. The value of ε is generally 4–5%.

The generator characteristic constant K_G [% MW/Hz], which is a constant that indicates how % of the rated output changes when the frequency changes by 1 Hz.

$$K_G = \frac{100}{f_0 - f_N} \tag{2.7}$$

It is expressed as the Eq. 2.6.

Similarly, as the frequency characteristics of the load, the power consumption of the load increases as the frequency increases, and the power consumption decreases as the frequency decreases. This is because, in the case of a rotating machine load included in the load, as the frequency increases, the number of revolutions of the motor also increases, and the power consumption increases accordingly. In addition, the power consumption of electric heating devices and lights often increases with the increase of frequency. There is a constant that represents how many MW the load fluctuates with respect to the fluctuation of this frequency. By the load characteristic constant K_L [% MW/Hz], and the frequency fluctuation of the entire power system called the system constant K [%MW/Hz]. The system has some degree of self-controllability because the generator output and load frequency increase/decrease in response to frequency fluctuation. As shown in Fig. 2.5, the point becomes a stable point where the state settles at the intersection of the two lines that show the frequency characteristics on the power generation side and the load side.

$$K_G = \frac{100 \times 100}{\varepsilon \times f_N} \tag{2.8}$$

$$K = K_G + K_L \tag{2.9}$$

Considering the whole power network where multiple generators are connected, the power ΔP [MW] of the difference between the total generation amount and the total demand amount and the fluctuation amount Δf [Hz] of the grid frequency have the relationship as Eq. 2.10.

$$\Delta P = K \cdot \Delta f \tag{2.10}$$

That is, as the value of the system constant is larger, the frequency fluctuation due to the supply–demand deviation generated in the system is smaller and the frequency is less likely to fluctuate.

LFC control is a control method that detects the frequency fluctuation amount due to load fluctuation, the interconnection power fluctuation amount, etc., and uses the power system frequency at steady state and the power flow of the interconnection line as a reference value. This is done for the fringe component. The fringe component is load fluctuation that occurs in a short cycle, and it is difficult to predict. Therefore, LFC is required to control the generator in real time against the load which changes from moment to moment.

In general practice, each power company monitors and controls the grid in each region. The grids are connected by transmission lines called interconnections. By connecting the grids, the scale of the grid is increased and the fluctuation of the frequency is reduced. In addition, since the generation reserve of each other can be used, when the amount of power generation in its own area is insufficient, it is possible to maintain the balance between supply and demand by using generators in other areas. A schematic diagram is shown in Fig. 2.6 from [3].

Also, when the balance between the amount of power generation and the demand amount in the area is unstable, power flows to the other area through the interconnection line. This is called interconnection current flow. Therefore, by measuring this tidal current, it is possible to judge how well the supply and demand in the area are balanced.

Because LFC control is performed, it is difficult to measure load fluctuation in real time. Therefore, in order to calculate load fluctuation in the area equivalently, frequency fluctuation and interconnection flow fluctuation are measured. As a result, it can be said that it is possible to maintain the grid frequency and the value of the interconnect flow at the reference value in order to maintain the load fluctuation amount at the reference value.

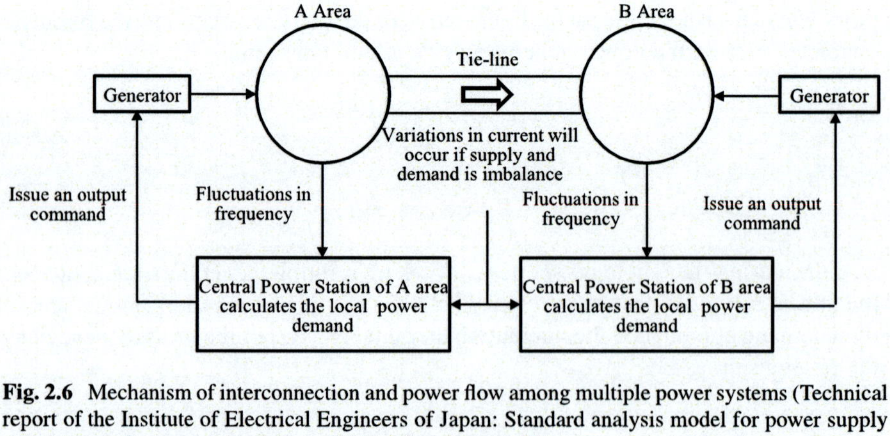

Fig. 2.6 Mechanism of interconnection and power flow among multiple power systems (Technical report of the Institute of Electrical Engineers of Japan: Standard analysis model for power supply and demand, frequency simulation (2016), taken from p. 75)

The control method of LFC is described. Two methods are also used to control the frequency, low frequency control (Flat Frequency Control: FFC) and Tie line power frequency Bias Control (TBC).

The FFC method is a control method that detects the deviation Δf of the grid frequency from the reference frequency, controls the generator output, and keeps only the grid frequency at the specified value. This is an effective method in a single system. But in a interconnection system, since this system does not control interconnection power flow, the power flow is disturbed by frequency control and affects other system areas.

The TBC method simultaneously detects the system frequency deviation Δf and the interconnection line power flow deviation, ΔP_T, and based on these, determines the supply/demand imbalance ΔP in its own area. This ΔP is called the (AR: Area Requirement) as the power required for the area. This is a method to control the generator output in order to bring this AR close to zero. Unlike the FFC method, it is possible to suppress the influence of interconnecting power flow on other system areas in the interconnection system.

The supply–demand imbalances ΔP_A and ΔP_B of grid areas A and B are calculated by Eqs. 2.11 and 2.12.

$$\begin{cases} \Delta P_A = K_A \Delta f_A + \Delta P_T \\ \Delta P_B = K_B \Delta f_B - \Delta P_T \end{cases} \quad \text{(2.11 and 2.12)}$$

Here, K_A and K_B are system constants of each system area, and Δf_A and Δf_B are frequency fluctuations of each area, and ΔP_T is the fluctuation of interconnection line flow from A area to B area. Because the unbalance components cannot be measured, the unbalance components ΔP_A and ΔP_B can be obtained from the measurable system constants K_A and K_B, the system frequencies Δf_A and Δf_B, and the interconnect flow tidal fluctuation component ΔP_T.

In LFC, grid frequency is maintained by giving control commands to the target generator. The response range of LFC is about $\pm$ 5% of the rated output of the generator.

2.2.2 Economic Dispatch Control EDC

Among the control for each generator from the central power supply command station of the power system, the economic load distribution control EDC (Economic Dispatch Control) is a control mainly in charge of the sustain component of load fluctuation of about several tens of minutes or more. In this time domain, it is control from the viewpoint of how it is economical to share loads with a large number of generators, instead of dealing with emergency measures once a minute and a second.

The fluctuation of demand more than several tens of minutes is larger than the fluctuation of every second and every minute, but it is within the forecastable area because the demand fluctuation is temporally gradual. Based on the demand fluctuation forecast up to several tens of minutes ahead, the generators with different economics due to fuel cost etc. are normally re-optimized in a cycle of about 5 min, and the power generation output for each EDC compatible generator is a control that outputs an up and down signal.

In addition, thermal power generators are generally more efficient when operated at high power. So the optimal operation/stop state of the generator is planned for the predicted demand. However, because it is control based on prediction, it is inevitable that a difference with the actual momentary fluctuation occurs. Therefore, the difference in time order, such as every minute to several tens of minutes, is corrected and controlled by the LFC. EDC usually updates the control output to each power plant by computing and processing predictions and corrections at time intervals of around 5 min.

The procedure of EDC arithmetic processing is shown in Fig. 2.7. First, we calculate the total allocation of EDC control from the demand fluctuation forecast and the power generation plan several tens of minutes ahead of that time. Next, the optimal allocation calculation as shown in Fig. 2.8 is performed so that the total allocation amount is most economical for each EDC generator. And the control signal to each EDC target generator is output based on the result of the optimal allocation calculation. This series of EDC calculation cycles is 5 min as a typical value.

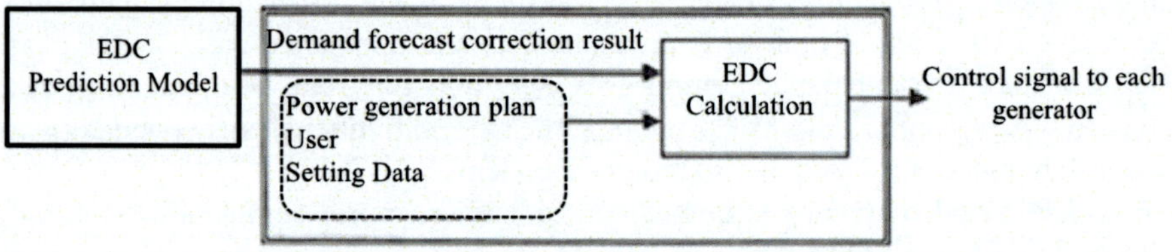

Fig. 2.7 EDC control procedure (Technical report of the Institute of Electrical Engineers of Japan: Reference analysis model for power supply and demand, frequency simulation (2016), taken from p. 83)

Fig. 2.8 Operation flow of EDC optimal load distribution

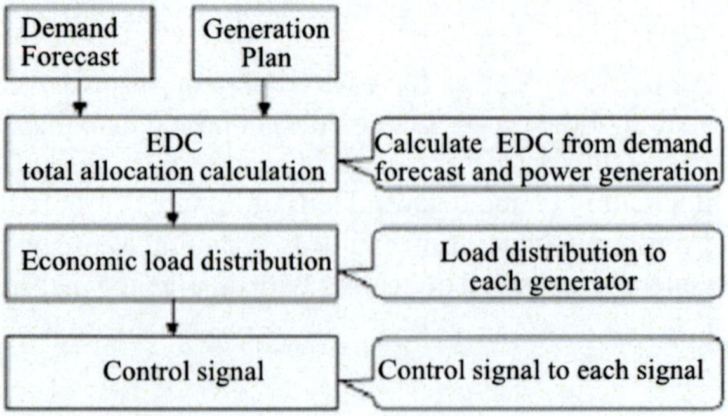

An example of EDC optimal calculation is outlined below. The demand fluctuation amount shared by EDC in the control cycle of EDC is expressed as Eq. 2.13 using the generator plan output and the demand forecast 5 min ahead.

$$P_{EDC} = L_C - (P_S + P_C + P_0) \tag{2.13}$$

Here, P_{EDC} is EDC share, L_C is forecast demand value of 5 min ahead, P_S. is total planned output of generators subject to EDC. P_C is total planned outputs such as base power sources not subject to EDC, P_0. is a planned value of the electric power which uses interconnection from other areas and interchanges.

The equal λ method used in the optimization calculation used in the economic load distribution process will be described. Generally, the fuel characteristics of a thermal power generator are said to be approximately represented by a quadratic function as shown in Fig. 2.9. For the EDC contribution P_{EDC}, the allocation P_{Gi} to each generator is an optimization problem that minimizes the objective function in Eq. 2.14.

$$f(P_G) = \sum_{i=1}^{n} f_i(P_{Gi}) = \sum_{i=1}^{n} (a_i + b_i P_{Gi} + c_i P_{Gi}^2) \tag{2.14}$$

Here, f_{G_P} is an objective function, the total fuel cost at the target time, and P_G is the total of the generator output. $f_i(P_{Gi})$ is the fuel cost function of generator i, P_{Gi} is the output of generator i, n is the number of EDC operating generators, and a, b, c are the coefficients of the quadratic curve that approximates the fuel cost characteristics.

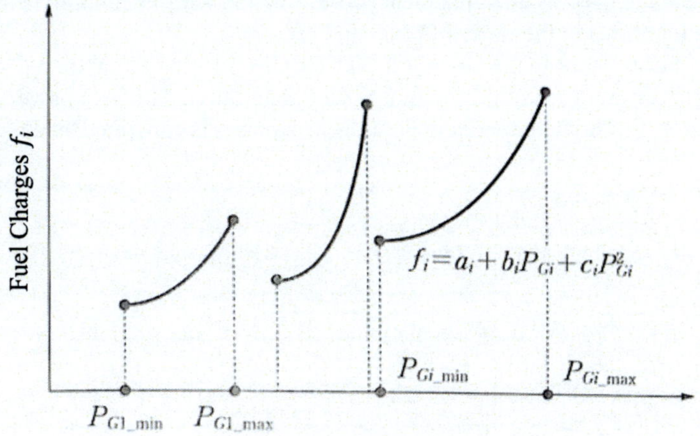

Fig. 2.9 Approximate by a quadratic curve with different fuel consumptn characteristics for each thermal power generator

The state in which the objective function of the Eq. 2.14 is the smallest is the most economical allocation.

This optimization problem is generally solved using Lagrange's undetermined multiplier method. This method is used for the EDC control simulation of power systems including virtual power plants, which is shown in Chap. 7 of this book.

To calculate the undetermined multiplier method with the Lagrange multiplier as λ, the Eq. 2.15 will be involved.

$$\lambda = \left(\sum_{i=1}^{n} \frac{b_i}{c_i} + 2P_{EDC} \right) \Big/ \sum_{i=1}^{n} \frac{1}{c_i} \tag{2.15}$$

Here, λ is called equal fuel cost, and when this λ is equal for all generators, the most economical power distribution can be realized. This is called the law of equal incremental fuel cost, or the equal λ method.

Since λ is determined by the size of P_{EDC}, λ increases as the load increases. The output of each of the two generators is G_1 and G_2, and the output change example is shown in Fig. 2.10. When $\lambda = 2.5$, the generator output is $P_{G1\_\lambda 1}$ and $P_{G2\_\lambda 1}$. When the demand fluctuates and P_{EDC} increases to $\lambda = 2.6$, the generator outputs become $P_{G1\_\lambda 2}$ and $P_{G2\_\lambda 2}$.

Figure 2.11 shows the unit price and adjustment power of various generators. Coal, petroleum, and liquefied natural gas (LNG) are used as fuel for thermal power generators, but each electric power company adjusts the generators to be operated according to the current fuel cost fluctuation. However, since coal is the least expensive, coal-fired power generators are used as base power sources that are constantly used without fluctuating output. On the other hand, thermal power generators using these fuels are used as an adjustment power to cope with fluctuating demand, since the cost increases in the order of LNG and oil.

In addition, there is the so-called GTCC (Gas Turbine Combined Cycle) as the latest one different from old thermal power generation. This is a power generation system in which high temperature gas is blown, power is generated by the gas turbine,

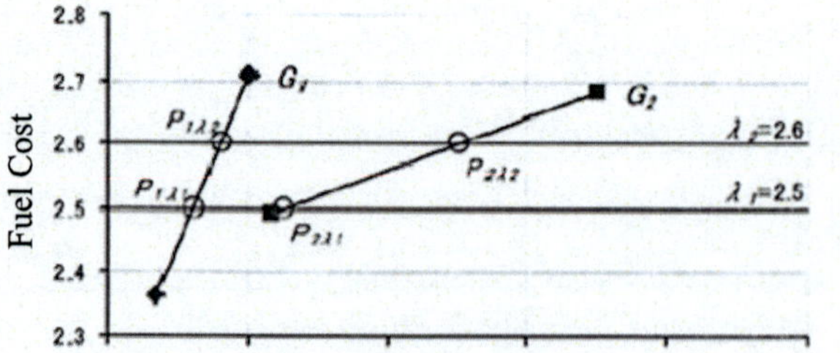

Fig. 2.10 Fuel cost change and output change of each generator (Technical report of the Institute of Electrical Engineers of Japan: Standard analysis model for power supply and demand, frequency simulation (2016), quoted from p. 84)

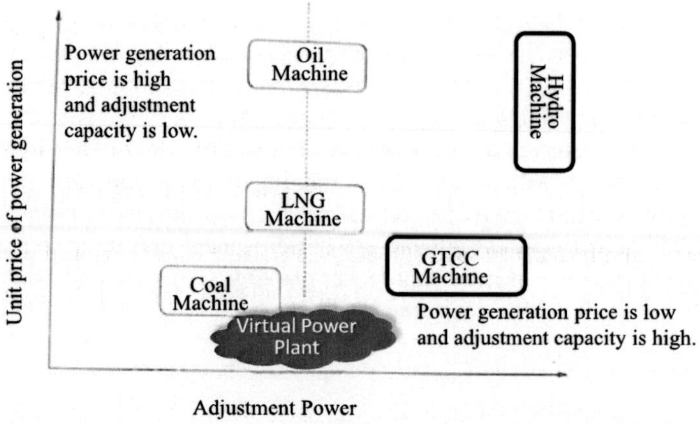

Fig. 2.11 Unit generation costs and adjustment power of each generator at Central Power Supply Control Station (Technical report of the Institute of Electrical Engineers of Japan: Standard analysis model for power supply and demand frequency simulation (cited from 2016, p. 98)

and then steam is generated by the heat of the still high temperature exhaust gas, and the steam turbine is also turned. It is a highly efficient power generation method compared to LNG machines.

Based on above, as an example, the outline of calculating the load distribution to each generator whose total fuel ratio is minimized by approximating the simple output fuel cost curve of the thermal power plant group with a quadratic curve is shown. As for other generators, base load generators such as nuclear power generation that hold a fixed power generation output are different. But in practice, when comparing EDC targets such as hydroelectric generators, optimization of output contrast operation cost it needs to be included in the calculation.

Nowadays, the EDC controlled generators from the central power supply command center are large-scale generators with existing technology. However, in the future, it can be expected that technology development will be possible to include the virtual power plants described in this book as EDC controlled power generation.

In future, if such a reliable and accurate virtual power plant system could be develop and there is an economic merit, it seems that the central power dispatching center shifted from the conventional thermal power generator to the virtual power plant. There is a possibility of being able to target various EDC control commands. In that case, the optimal calculation of EDC includes the economic efficiency of the virtual power plant as an evaluation value equivalent to the fuel cost and includes it in the optimal calculation.

It is important that the virtual power plant, especially the virtual power plant by FastADR negawatt of the building multi-type air-conditioner group used in this book, does not incur the capital investment and maintenance cost for that. It is necessary to reflect the economics including the construction investment recovery of large-scale thermal power plants, the cost of maintaining the annual operation status, and the real estate and other fixed cost repairs. Regarding fuel cost, negawat virtual power plant of building multi-type air-conditioning group is not necessary, but the corresponding cost is not zero, mainly for incentives for customers who cooperated with building multi-type air-conditioning through service providers.

However, the inclusion of negawatt virtual power plants as EDC targets for large thermal power plants with low annual availability and CO_2 emissions during operation seems to be promising in the future. For that purpose, it will be the key to build up the high-speed, accurate, reliable power adjustment performance equivalent to the conventional large-scale thermal power plant by accumulating basic technology of IT and AI which will be described in this book.

References

1. P. Kundur, *Power System Stability and Control* (McGraw-Hill, Inc., 1994)
2. Technical report of the Inst. of Electrical Engineers of Japan, *Constant and Emergency Load Frequency Control in the Power System*, No. 869 (2002)
3. Technical report of the Inst. of Electrical Engineers of Japan, *Standard Analysis Model for Power Supply and Demand/Frequency Simulation*, No. 1386 (2016)

Chapter 3
Virtual Power Plant System

3.1 Virtual Power Plant Concept

3.1.1 Power Demand and Negawatt

As described in the proceeding chapters, in present power system, the demand and supply are controlled only by generation side, according to the fluctuation of the power consumption on the demand side. However, if the number of renewable power generation facilities such as solar power generation and wind power generation increases, the proportion of the amount of supply that can be controlled by governor control or LFC control described in the Chap. 2 decreases relatively and control according to the demand side becomes difficult.

Then, the method of controlling the electric energy on the demand side is considered. This is called Demand Response (DR). In order to balance supply and demand, suppressing consumption on the demand side is equivalent to increasing power generation. Therefore, even if the power generation of the whole system is sharply reduced due to a sudden change of weather or an accident, the power consumption on the demand side can be suppressed and the balance of supply and demand can be maintained before a large-scale blackout occurs.

Demand response uses IoT communication technology to control the power consumption on the demand side according to the status of the power system instantly. There is a "smart grid" that combines conventional power control technology and IoT communication technology. This is a next-generation power network, with the goal of performing power supply systems with high efficiency, high quality, and high reliability.

Although there is often misunderstanding, the demand response is different from energy saving in which the demand side uses less power consumption as much as possible. This is because demand response not only suppresses power consumption, but also performs control to balance power consumption by increasing power generation when the amount of power generation is too large. In addition, incentives are

C. Ninagawa, *Virtual Power Plant System Integration Technology*, Power Systems,
https://doi.org/10.1007/978-981-16-6148-8_3

also produced corresponding to the power consumption that fluctuates due to the power consumption control (an incentive for supplying power).

Thus, for balance between supply and demand, increment in virtual power generation by demand-side power control is called negawatt; refers to the amount of power reduced as a result of consumers refraining from their original intended power consumption levels, in response to the requirement by power utility and/or market. Currently, storage batteries, air-conditioners in office buildings, large-scale water heaters in hotels and hospitals, etc. are considered as the sources of negawatt generation. In other words, various loads are targeted for which the effect of the reduction of power consumption does not affect work and life instantaneously. In the case of storage batteries and water heaters, it is sufficient to store the amount of power and hot water required by a predetermined time, so if the pace of charging and hot water supply is delayed for a short period of time, the working environment and living things will not disturb immediately. In addition, it is possible to obtain the negawatt by suppressing the air-conditioners power in office building for about 10 min without disturbing the human's comfortable zone.

On the other hand, by increasing the power demand, reduction the amount of virtual power generation is called positive watt; refers to absorb the power by additional the power demand when power system has enough power (more than actual demand). Batteries and water heaters, and other electric vehicles are also considered as sources of positive watt. In addition, increasing the power consumption of the air-conditioner is practically difficult because pre-cooling and pre-heating are performed. However, storage batteries and heat pump water heaters can be used to store energy by means of positive watt, so they are useful as a positive power generation source.

3.1.2 Customer Side Virtual Power Plant Services

Controlling the power consumption of the customer by the demand response, described in the previous section, is the key to the concept of virtual power plant (VPP). Figure 3.1 shows a schematic image of the virtual power plant. In virtual power plant, each facility such as storage battery, water heater and air-conditioners of the customer can be controlled, monitored and managed by distribution of control command from the central power supply command center, and the aggregation of generated negawatts is performed by a middle party (business operator) called aggregator.

There are two types of aggregators: resource aggregators and aggregation coordinators. A resource aggregator directly manages and controls each facility (energy resource). On the other hand, the aggregation coordinator is to trade the amount of power controlled by the resource aggregator with general transmission and distribution operators and retail electric utilities. By creating such a virtual power plant, it is

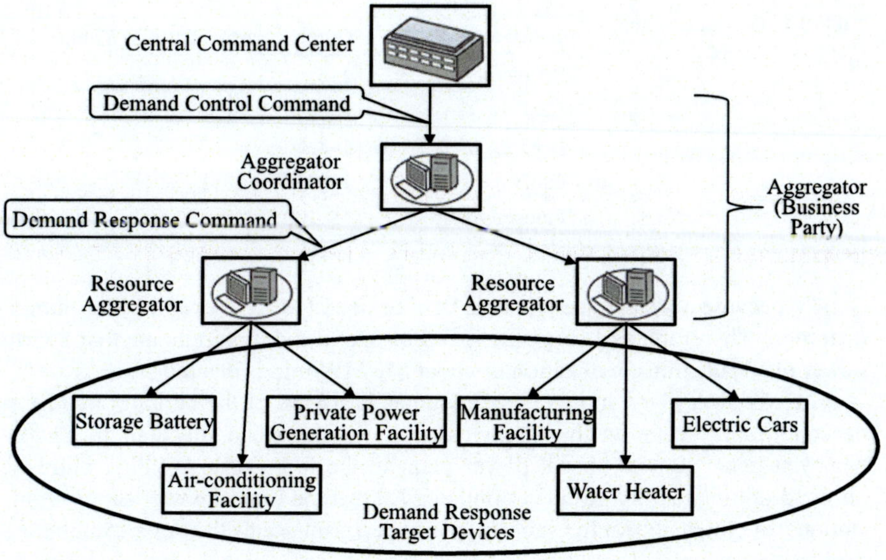

Fig. 3.1 System image of virtual power plant

possible to demonstrate the ability to adjust the frequency to the grid, achieve reduction of electricity charges, and provide services such as avoidance for photovoltaic power generation output suppression.

The negawatt trading is being conducted as a business utilizing virtual power plants. Negawatt trading is a trade that generates power based on the demand response control command from the aggregator based on a prior contract with the aggregator, and produces the incentives according to the reduction of the power consumption. Demand control that generates compensation according to such fluctuations in power is called an incentive type. In the negawatt transaction flow, the aggregator receives the output control command from the central power supply command center, and issues the demand response control command to the customer's equipment under its control. Aggregators aggregate the amount of power that fluctuates due to the demand response control command, receive the incentive according to the fluctuation from the power company, and the customer is paid the incentive.

In order to virtual power plant to cooperate with the thermal power generator for supply and demand adjustment, it is important that it can be responsive from the central power supply control station. Naturally, the response speed to the demand response command differs depending on the load used as the control target, but in order to use the virtual power plant as the adjustment power for frequency control, the response speed equivalent to the actual power plant such as the thermal power plant is required. Table 3.1 shows the output change rate of each power plant as in [1]. Since GTCC has a median value of around 5% MW/min, the output change rate is faster than other thermal power generators, and especially the pumped storage hydro generator is extremely fast. In addition, it can be said that frequency adjustment

Table 3.1 Generator output change rate (rated ratio)

Generator type	Output change rate (% MW/min)
Petroleum	1–8
Coal	1–5
LNG	1–8
GTCC	1–10
Pumped-hydro	15–100

is easy because variable speed steam can be operated LFC even during pumping operation. Therefore, as a response speed of one of the requirements that a virtual power plant functions, realization of about 5% MW/min is desired.

Virtual power plant, in this book, assumes FastADR of the building multi-type-air-conditioner group as the main resource. As described in detail in Chap. 7, negawatt generation of virtual power plant in the large-scale building multi-type air-conditioning facility group is simulated [2] that the central power supply control station can obtain almost the same response performance as the target generator for grid frequency control.

Also, Fig. 3.2a shows the response characteristics of a typical gas turbine thermal power plant for supply and demand adjustment with respect to the Economic Dispatch Control (EDC) control command from the central power supply control center.

Figure 3.2b shows an example of the negawatt production response when integrating multiple FastADR response waveforms of various building multi-type air-conditioners [3]. Here are the results of three simulations in which ADR power response time-series data of 100 building multi-type air-conditioners were combined intensively and regarded as simultaneous execution. In this example, about 2 min response time is obtained, which is almost the same as a gas turbine thermal power generator. Since the number of units is as small as 100 units, the negative absolute value of negawatt generation is only about 0.4 MW, but if high-speed communication control technology can be established to carry out this accurately to about 100 units, the response characteristics similar to existing thermal power generators it suggests the possibility of being obtained.

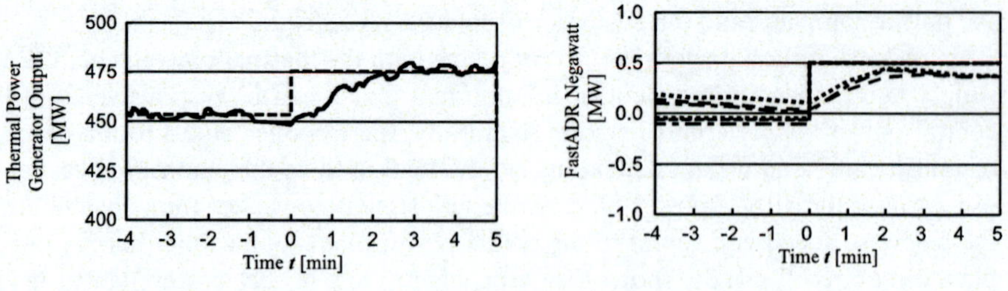

Fig. 3.2 Building multi-type air-conditioning negawatt response characteristics to central power supply control center [2]

In this way, large-scale and high-speed communication is the main point for precise and delay-free aggregation control of FastADR negawatts for a large number of building multi-type air-conditioners installed in a large number of dispersed office buildings. In the next section, we will introduce communication standards assumed for such large-scale and high-speed aggregation control of FastADR in context of virtual power plants. Here, only the outline of each communication standard is discussed, the communication data transmission performance at the time of aggregation is described in Chap. 7, and the specific communication software development technology is described in detail in Chap. 8.

3.2 Communication Systems in Virtual Power Plant

3.2.1 OpenADR Indirect Control Communication

OpenADR (open automated demand response) [4] specifies the automation of demand response, that is, the data exchange protocol for automatic demand response (ADR). Therefore, communication services (machine registration, demand response, event notification, device status report, demand response participation/non-participation notification, etc.) required for the implementation of demand response are premised on direct interaction between computers (machine to machine: M2M). Sequences, data payloads, and communication protocols are defined. Also, it is open; that is public and anyone can access it.

The beginning of OpenADR is the 2002 California state power crisis. In the wake of the crisis, the Lawrence Berkeley National Laboratory in the United States began researching automatic demand response, and after developing a demonstration test and commercialization of the California government and a company jointly with several companies. A summary of the results was released in April 2009 as the OpenADR 1.0 communication specification [5].

In the same year, it was selected as one of the standards in the Smart Grid Interoperability Panel (SGIP) established by the NIST (National Institute of Standards and Technology Research Institute), and was adopted in the PAP (Preferred Action Plan).

UCAIug (the power company's international user group) developed an OpenADR 1.0 system requirements specification that conforms the locally colored OpenADR communication specifications to other international standard models, and approved it in September 2010.

OASIS (an International non-profit Organization Promoting Development of Open Standards), which was considering the standardization of ADR at the request of SGIP, based on the OpenADR 1.0 system requirements of UCAIug and the examination results of NAESB (North American Energy Standards Committee), etc. In February 2012, EI 1.0 (Energy Interoperation 1.0) [6] was released a new international standard. Although this is a vast standard aiming at interoperability not only

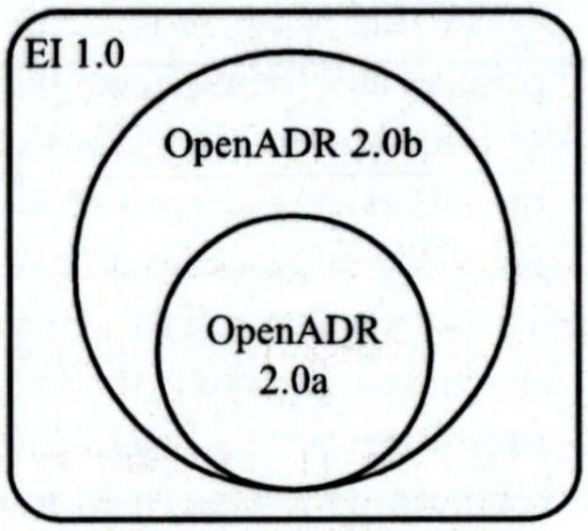

Fig. 3.3 Relationship between OpenADR 2.0 Profiles and EI 1.0

for ADR but for the whole power and energy field, there are also aspects that are insufficient for practical use, and a more detailed standard was desired.

Based on the specification of EI 1.0, OpenADR 2.0 was formulated as a standard that is limited to ADR and considered for implementation. Currently, management, certification, and promotion are being conducted by the OpenADR Alliance based in California, USA, which was established in 2010.

OpenADR consists of two versions, OpenADR 2.0a and OpenADR 2.0b. Standards that are the part of such higher-level standards are called profiles. So, they may be referred to as profile A and profile B, respectively. The relationship between each profile and EI 1.0 is nested as shown in Fig. 3.3, that is, the upper profile includes the lower profiles.

OpenADR 2.0a profile released in August 2012. It targets control of relatively simple devices, and supports only partial communication of demand response events.

The profile released in July 2013 is OpenADR 2.0b, and as of November 2015, revision 1.1 is also released. OpenADR 2.0b is aimed at full-scale automatic demand response service. And in addition to communication of required communication service and demand response event that also considers cooperation with business partners such as aggregators and clients on the customer side. Resource registration and status also includes in OpenADR 2.0b. Initially, OpenADR 2.0c was also planned, but as a result of discussions with electric power companies, it was judged that OpenADR 2.0b is sufficient.

OpenADR 2.0b is currently aiming for international standardization in the IEC (International Electrotechnical Commission). Public Specification (PAS) was approved as IEC/PAS 62746-10-1 in February 2014, and in May 2018, a vote was called about the international standard draft of the PC 118 (project committee dealing with user interface of smart grid). OpenADR 2.0b was passed by a large number votes. Therefore, it is expected to be standardized internationally around June 2019 if it goes well.

Figure 3.4 shows the adoption flow of OpenADR standard, specific for Japan. Also, as mentioned above, since OpenADR 2.0 is the mainstream at present, we will refer to OpenADR 2.0 when we simply call it OpenADR from this point on.

Participants in an automatic demand response system can be divided into two major groups: demand response event issuers and consumers who receive them. Originally, it was also possible to support data communication to consumers, but it

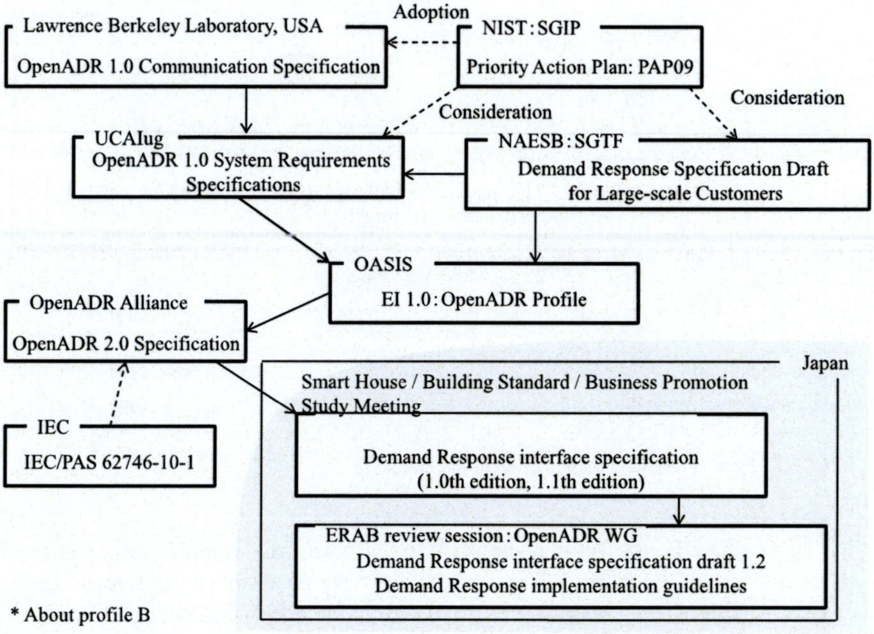

Fig. 3.4 Flow of adopting OpenADR 2.0

is better suited to the upper method of clustering on a large scale and aggregating automatic demand responses (aggregator). Here, the demand response event issuer corresponds to a power transmission and distribution operator, a system operator, a retailer, and so on. On the other hand, consumers include various devices, and their sizes also vary. For example, factories, large-scale buildings, power plants, aggregators (a bundle of multiple small customers), micro grids, and Regional Energy Management Systems (CEMS) can be consider as large-scale customers. And not only the Building Energy Management System (BEMS)/Home Energy Management System (HEMS), but, also individual air-conditioners, storage batteries, and generators can be considered for small-scale customers.

In OpenADR, these participants are called actors, and adopt an information transmission model that relays demand response events from upstream actors to downstream actors and relays intermediate actors in multiple stages. Specifically, when a demand response event is issued by an upstream actor, the demand response event is first transmitted to the child actors belonging to it through so-called client/server communication. Next, the demand response event is similarly transmitted to the grandchild actor who belongs to the child actor. By repeating this process, the demand response event is transmitted from upstream to downstream. As shown in Fig. 3.5, the middle actors are both the receiver and the sender of the demand response event, and communication between actors that are not directly related does not occur.

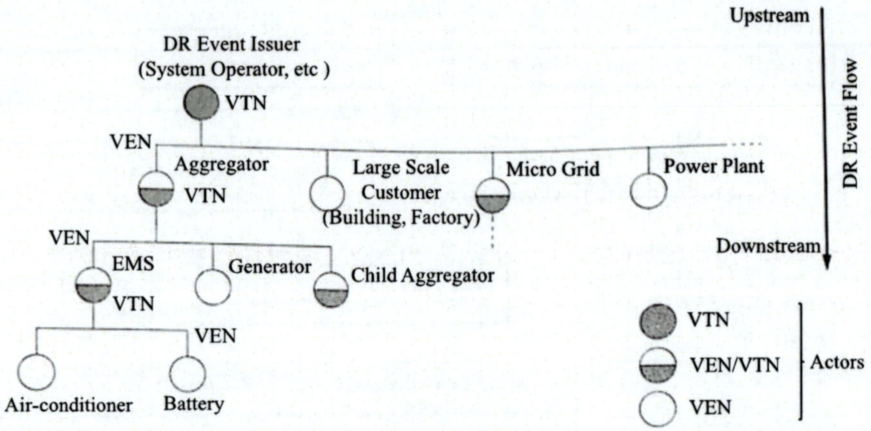

Fig. 3.5 Flow of demand response event in OpenADR

In OpenADR, the actor responsible for sending the demand response event is called (Virtual Top Node VTN), and the actor which receiver the demand response event is called virtual end node (Virtual End Node VEN). However, VTN and VEN are not fixed, and an actor can be either VEN or VTN.

3.2.2 *IEC 61850 Direct Control Communication*

This subsection describes international communication standards for direct control of the devices that make up a virtual power plant from the resource aggregator of the virtual power plant through the energy resource controller of the customer.

Direct control communication handles specific information unlike the OpenADR which communicates from the top aggregation coordinator to the resource aggregator described in the previous section. In order to measure the sensor value of the target equipment or adjust the power directly, it is necessary to send a detailed signal according to the actual equipment. The communication and control device on the customer side convert the control signal into a device-specific monitoring control signal, communication is standardized, so that, it does not depend on the equipment type, manufacturer, and model.

The state of standardization at the time of the writing this book seems to be that OpenADR becomes the de facto standard for communication with the upstream system. In this book, two communication standards are taken as communication on the downstream side of the virtual power plant system, that is, on the customer side. That is, the IEC 61850 communication standard [7], which has been standardized in the field of power transmission and distribution, and the IEEE 1888 communication standard [8], which has been standardized internationally for power monitoring and control, are treated as representative examples.

This subsection describes the IEC 61850 communication standard series. The IEC is the International Electro-technical Commission for Power Systems, and this communication standard is the series of communication standards for the automation of power facilities formulated by the technical committee TC57.

Initially, when communication control is performed by combining power devices supplied from many vendors in a power facility, expensive communication converters are required for each vendor. Providing interoperability for control communication between devices of different vendors is an advantage for users. Therefore, interoperability is a common goal of utilities, equipment vendors and standardization organizations. The IEC project group, which consists of about 60 members of different countries, started work in 1995 with three IEC working groups.

Originally, it began with the standardization of communication for automation of substations in the power system, but the scope of coverage has gradually expanded. In recent years, it has been widely targeted, including distributed power sources such as wind power generation facilities and solar power generation facilities. As a result, it came to be recognized as a standard that can handle various electric power facilities in a unified manner.

Thus, at the beginning we started to automate the control communication between devices in the substation, but the scope of application has been gradually expanded. And this standard has been accepted in general by smart grids, because various control information models including distributed power sources such as wind power generation facilities and solar power generation facilities have been defined.

Recently, a framework concept called OS4ES [9] was proposed in Europe as a demand-side aggregation management system. It is a management system that queries and combines distributed energy resources such as solar power, wind power and demand response distributed in a wide area with grid users in real time. There, it is necessary to unify and model various distributed resources into an information model and systematize it as a database. Here too, the abstracted control information model is assumed for IEC 61850 communication standard series.

Nowadays, this communication standard series has become a standard communication standard as an information model that uniformly represents a power system, rather than for power equipment in a power network.

We will briefly explain the outline of this communication standard series as basic knowledge for understanding the entire communication system of the virtual power plant. Figure 3.6 conceptually shows the positioning of the individual standards that make up this series of communication standard. That is, it is systematically constructed from many individual standards regarding standardization of information representation of controlled objects, standardization of information exchange, and standardization of information exchange execution.

In this communication standard series, standards such as IEC 61850-7-2, IEC 61850-7-3, IEC 61850-7-4, and IEC 61850-8-1 are mentioned as individual standards related to direct control of virtual power plants. This subsection will only briefly describe their position, and we will describe in detail the specific communication system development method later in this book.

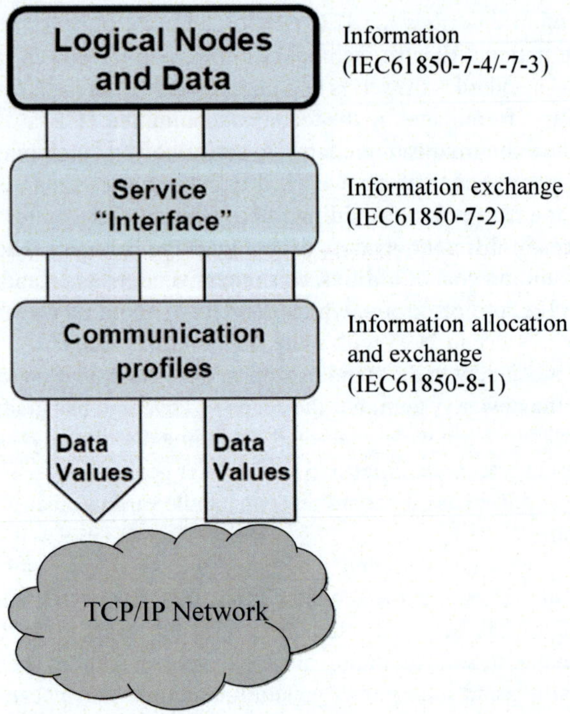

Fig. 3.6 Positioning of individual standards that make up the IEC 61850 communication standard series

IEC 61850-7-2 defines services for exchanging information on the functions of various power installations, such as control, reporting, acquisition and configuration. Within the scope of this standard, these pieces of information are expressed in an abstract form, and are separated from the communication protocol and the specific communication service that realize it.

In IEC 61850, function is refers to task which performed by the substation automation system. Generally, a function consists of subparts called logical nodes, which exchange data with each other. By definition, only logical nodes exchange data and, therefore, a function that exchanges data with other functions must have at least one logical node. As a consequence, only data contained in logical nodes can be exchanged in the context of the IEC 61850 series.

IEC 61850-7-3 defines a common basic information model, so, that it does not depend on the manufacturer or model of control or measurement. IEC 61850-7-4 abstractly defines the information items of automated models of power equipment, for example, circuit breakers and protective elements. The individual standards are abstractly modeled as common functions, such as the open/close state of the circuit breaker and the setting of protection functions, etc., for monitoring and controlling the power equipment, and control information is exchanged even between different manufacturers.

IEC 61850-8-1 specifies specific means for communicating information between power devices and a method for serializing and exchanging information. In other words, it specifies the correspondence to the power equipment application layer and the mapping correspondence to the lower layer standard in concrete communication.

One of the aims of this power equipment IEC 61850 communication standard series is the support of the power equipment automation function by communication. The automation function of the power equipment is, for example, reading of sensor measurement data for current measurement, such as transmission of a shutoff operation signal for overcurrent protection. In addition to measurement data and blocking signals, engineering and equipment configuration, monitoring and control, control center communication, time synchronization, etc., functions such as measurement, status monitoring, and asset management are also supported. Although many functions are implemented in power devices, implementing a single function in a single power device or implementing multiple functions in a single power device is another feature. It can also be hosted on a power device. A power device communicates with the functions of other power devices by means of an information exchange mechanism.

The information exchange mechanism of the IEC 61850 communication standard series is based on a well-defined information model. These information models and modeling methods are the core of this communication standard series. The philosophy of this communication standard series is to center on the abstract common information model. This defines the method of information exchange separated from the actual communication lower layer implementation, and makes it possible to express using the concept of virtualization. Also, although different in the manufacturer and model of the device, only the details necessary to achieve the interoperability by communication between them are defined.

The abstraction that is the philosophy of this communication standard series is explained below. As defined in the specific standard IEC 61850-5, the basic approach of this series of standards starts with breaking down application functions for exchanging the information into smallest part. These smallest parts are called Logical Nodes, which are the terms that characterize this communication standard. By standardizing the naming rules for logical node names, it is possible to identify the meaning regardless of various power equipment and manufacturers. For example, in the case of a circuit breaker (CB), systematic nomenclature is standardized, such that CB is a logical node named as XCBR. Thus, logic nodes are modeled and defined from a conceptual point of view by the individual standard IEC 61850-5.

By constructing a Logical Device from several logical nodes, it is possible to provide an abstract representation of the configuration of the entire power facility. Therefore, a logical device group that abstracts, these devices are defined as a direct control object, as a communication object, assuming that one of them is implemented in a power facility.

An example is shown in Fig. 3.7. The device is defined to consist of several logical devices and corresponds to the function of the actual device. In this example, logic

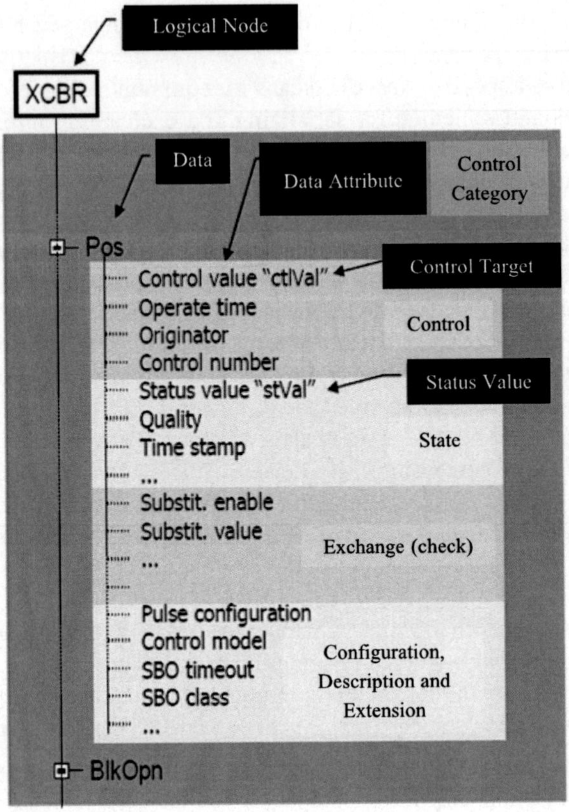

Fig. 3.7 Logical node and data structure taking a circuit breaker as an example (From IEC 61850-7-1 (2003), p. 19)

node XCBR abstractly represents the function of one circuit breaker on the right side of the power equipment. Each logical node has its own list of data objects based on its function. For example, information about CB position, which is a characteristic of the logical node, is a data attribute. Each data attribute contained in a logical node has its structure and its well-defined meaning. Attributes represented by data objects and their information are described in IEC 61850-5. Thus, it is exchanged by the clear rules and the requested services. The logical nodes and data objects included in the logical device are the key to realize the interoperability of direct control of the power equipment using this communication standard.

In IEC 61850-7-4, about 90 standard logic nodes are defined to cover the general application of a wide variety of power equipment. Interesting thing is that the group of applications related to protection make up almost half of the logical nodes, due to the high importance of protection for safe and reliable control of power equipment. However, in the present situation where the application is spread to renewable energy electric power facilities etc., the importance of the function regarding supervisory control is increasing, and this standard is in the process of being gradually supported.

Information held by logical nodes is represented by data objects. A logical node can hold several to the maximum of 30 data objects. Data objects may contain several to 20 or more data attributes. In this way, a logical node can include more than 100 individual information organized in a hierarchical structure, that is, a controlled object concept traditionally referred to as a point in the facility monitoring and control industry.

For example, in the logic node "circuit breaker XCBR", the data object named "Pos" regardless using the word "switch position" of the circuit breaker consists of several data attributes. For example, the control value of CB position is a data attribute of "Pos" data object; abbreviated as ctlVal. The value "Pos. ctlVal" taken by the CB position data object "Pos" of the logical node "XCBR" is standardized to the actual breaker position (intermediate state, off, on, failure state). Thus, the names of each control information are unified and standardized regardless of the manufacturer model of the device parts.

Services defined in IEC 61850-7-2 are called abstract services. A definition is defined to describe the action required of the service request receiver. Based on the functional requirements of IEC 61850-5, the meaning and behavior of the service and the parameters associated with the service request and response are defined in IEC 61850-7-2. The specific syntax (form), in particular the encoding of the messages carrying the service parameters of the service, and the way in which they pass through the network are defined in Specific Communication Service Mappings (SCSM) [10]. One specific standard IEC 61850-8-1 of SCSM is a detailed correspondence relationship (mapping) of services to MMS [11] (ISO 9506) and also TCP/IP and Ethernet as an example of a specific lower layer communication standard is standardized and defined. The SCSM mapping to other lower communication layers is defined in IEC 61850-9-2 and IEC 61850-3.

In addition, the individual standard IEC 61850-6 defines a unified description language for the power equipment. This language is called SCL (Substation Configuration Language) [12]. SCL can define control information of power equipment in a unified way, and a file format that can be used to exchange control setting information between communication devices is defined in a unified way. SCL can configure logic nodes to be suitable power service functions, map power devices to individual functions and devices, and define complete configurations and states that allow all communication.

As described above, in the IEC 61805 communication standard series, the control object of the power system is unified into a control information model, and the method to connect the access to it to the specific lower layer communication is organized. Figure 3.8 summarizes the contents described in this section as one conceptual diagram.

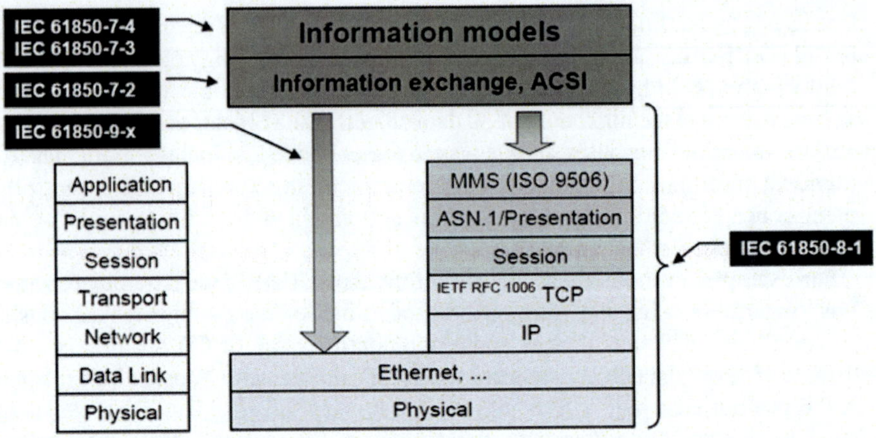

Fig. 3.8 Communication service mapping (SCSM) (Cited from IEC 61850-7-1 (2003), p. 22)

3.2.3 IEEE 1888 Communication for Customer Facility's Equipment

In this subsection, among the communication standards used for demand response communication on the customer side in the virtual power plant system, the communication standard IEEE1888 [13] on the customer equipment side considered as a candidate is described along with the IEC 61850 described in the previous subsection.

The IEEE (The Institute of Electrical and Electronics Engineers, Inc.) is an association of electrical and electronic engineering headquartered in the United States. Nowadays, it is appropriate to call it a conference of the whole world rather than a conference of the United States, and the members extend to countries around the world, and this group is the largest in the world. The name is electric and electronic, but covers a wide range of technology areas ranging from communication, information and control.

In addition to academic activities such as the publication of journals and the holding of international academic conferences, we are also carrying out activities to establish IEEE standards, which are standards for disseminating electricity and electronics and information, communication, and control technology to the world. The LAN communication standard (IEEE 802.3 Ethernet standard) and the Wi-Fi wireless communication standard (IEEE 802.11 series standard) are famous as examples of typical standards widely known in the world.

The IEEE 1888 standard, a communication protocol, was originally developed by the Green University of Tokyo Project (GUTP) [14]. In summer of 2011, when a power shortage occurred due to the Great East Japan Earthquake, 30% reduction in power consumption from the previous year during peak hours by visualizing the power consumption of the buildings on the 5 campuses of the University of Tokyo. Since then, we have been working on new evolution beyond wide-area building management. Figure 3.9 shows such ICT project demonstration experiment under Todai Green Project.

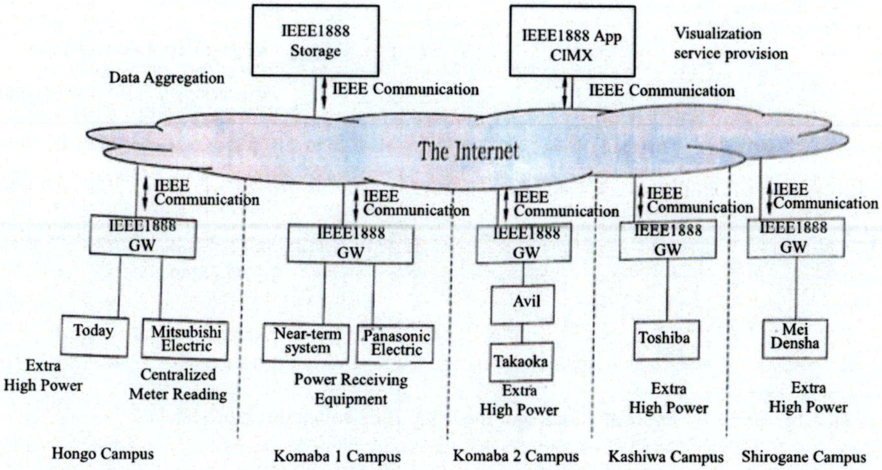

Fig. 3.9 The Todai Green ICT project demonstration experiment

This project was launched in 2008 as a university-industry-collaboration project of the University of Tokyo, and many demonstration experiments on monitoring and control of facility power consumption on the Internet using the University of Tokyo campus. Later, in March 2015, this standard was also certified as the international standard ISO/IEC/IEEE 18880.

This communication standard is based on HTTP (Hyper Text Transfer Protocol), which is widely used as a Web content transmission/reception protocol. Furthermore, the process control protocol for communication requests and communication responses that are intended to be used on the Internet combining a method called SOAP (Simple Object Access Protocol). It manages a pair of information request communication and information response communication. The intention is to use the existing TCP/IP network. Also, since it is HTTP-based, aiming to use existing technology smoothly using setting up a firewall when connecting to the Internet.

The feature of this communication standard is based not only on message exchange but on the basis of handling equipment information in the facility which is the object of supervisory control as a time series value. In addition, monitoring and control of electric power equipment is planned in business facilities such as remote office buildings on the premise of wide area IP network such as the Internet. It can be said to be a framework uses today's commonly used IoT (Internet of Things) technology in context of Internet control.

Figure 3.10 shows the basic configuration concept that abstracts the components of the IEEE 1888 communication system. The basic configuration consists of gateway GW (Gate Way), storage, APP (Application), and Registry. This basic configuration does not correspond to a specific device, but one that is responsible for storing facility monitoring information as a database, one that is responsible for converting to storage, individual control communication for facility devices, gateway GW, those

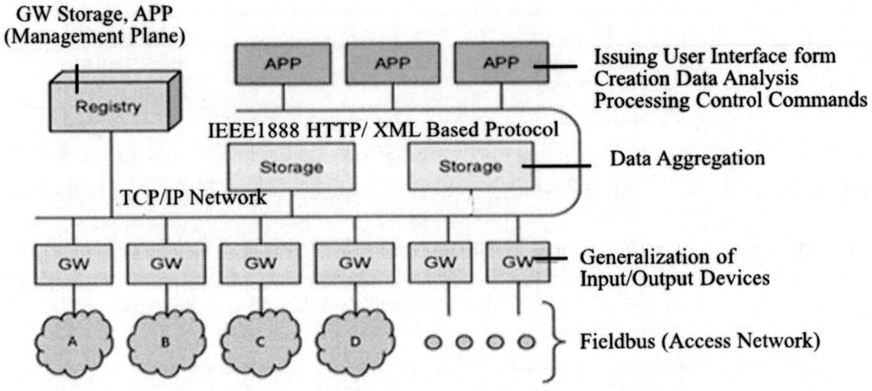

Fig. 3.10 Basic system configuration of the IEEE 1888 communication standard

supervisory control Running application software that makes useful use of information is considered to be APP. Physical devices may change the roles of GW, Storage, and APP depending on the situation, and may execute those functions in duplicate.

In the IEEE 1888 communication standard, three types of communication services such as FETCH, WRITE, and TRAP by Web service are defined among three parties like GW, Storage, and APP. The FETCH communication service is an information retrieval service that can be said to be the basic form of Web service of client server model. This is a communication service in which the application APP, a client application, makes a trigger and queries the storage using FETCH request from server, and reads out status information such as facility power.

In WRITE communication service the client creates a trigger, sends equipment information to the server, and writes it. The characteristics of Web service of the IEEE 1888 standard is the TRAP communication service. The TRAP communication service is communication in which the server creates a trigger and notifies the client of the occurrence of an event when a change in information on the server occurs. That is, it realizes the mechanism of "Server Push". The server Push operation is possible by devising the lower layer with other communication protocols for remote monitoring control. However, the feature of the TRAP communication service is that the client can register the server's desired event occurrence conditions in advance. In other words, it is possible to apply TRAP. This makes it possible to register different unique event occurrence conditions, that is, server push occurrence conditions, in the server for each client.

The application of this communication standard to the FastADR control system is described below. In the past, general demand response control such as power demand control of customer facilities has been performed monitoring control by time unit such as 30 min or 1 h. However, as a virtual power plant, in order to control the large-scale demand; electric power load must be controlled precisely. Therefore, it is necessary to have a communication method that can pass through the customer's firewall via the internet and can respond quickly.

In this communication standard, application of FastADR control [15–17] is sufficient for network connection only when communication is necessary. IEEE1888 standard for supervisory control has low communication cost by adopting RPC method that is Web service using HTTP and XML. The occurrence of a power limitation command is registered as the server push occurrence condition to the aggregation server in which the BEMS which is the client becomes a server in advance. When a demand suppression request to the BEMS is generated in the aggregation server, the server Push is notified of the demand suppression request to the BEMS.

Figure 3.11 shows the concept of FastADR control using IEEE 1888 standard for the downstream part of the virtual power plant, that is, the customer facilities group distributed over a wide area. This system is composed of two aggregation servers that are functionally distributed to the power company and aggregator hierarchically which performs FastADR control, and a group of Building Energy Management Systems (BEMS) in the customer premises. The aggregation server of the electric power company which arranges the demand response control of the entire customer facility group sends control information such as the demand suppression of electric power to the aggregation server of the aggregator which performs the demand response control for each customer facility group. In response to this, the aggregator's aggregation server divides the demand suppression request for each customer facility in the target customer facility group and distributes it to the BEMS in each customer

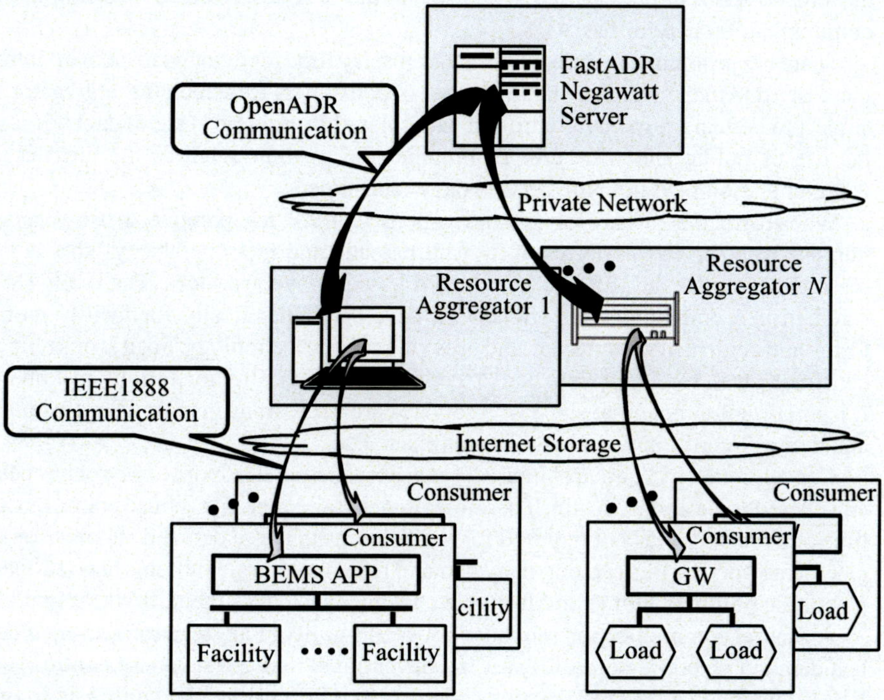

Fig. 3.11 Conceptual diagram of FastADR control for wide area building customer facilities

facility. The BEMS in the customer facility monitors and controls the building equipment according to the demand suppression request, and returns the control result to the aggregator's aggregation server. In most cases, the aggregator's aggregation server cannot establish a connection between BEMS and TCP/ IP in the customer premises due to the firewall installed at the network entrance of the customer's facility for securing the network security of the customer's facility.

As described above, in the general demand response control, the aggregator's aggregation server notifies the BEMS of customer's facility about power reduction value for a specified time slot, and waits for the response from the BEMS. In this demand response control, the network connection between the aggregator and the BEMS is often connected only when it is necessary to exchange information, in order to reduce the communication line cost and to secure the network security. For example, the BEMS in the customer facility periodically reads requests for suppression of the power demand from the aggregator's aggregation server, such as at one-minute intervals. For this reason, we need to send delays for the transfer the whole information.

On the other hand, the power company and the aggregator are connected by a dedicated line such as an optical line, and the information is exchanged, so that sufficient communication performance is secured. In addition, it has also been reported that the exchange of information between the electric power company and the aggregator was possible in about 2 s on average for the FastADR message, which included 100 customers. Also, in the OpenADR 2.0 standard, 4 s is described as an example for communication time of FastADR.

There is a problem that the communication delay time for demand response control increases due to the increase in the number of consumers for which the aggregator is subject to demand response control. In particular, this problem is significant because the use of public and wide area communication medium such as the internet as communication medium with service target consumers.

When using the internet for demand response control, it is practical to use general-purpose web services in terms of firewall passage and interoperability. That is the reason to consider the IEEE 1888 standard based on web services. The IEEE 1888 standard is a communication framework developed specifically for facility monitoring and control by the internet, and uses web service technology. So, it is possible to construct a web service system with excellent interoperability for various customers. It is important to note that since web services are used, transmission time guarantee and delivery confirmation are not accomplished.

Conventionally, the control interval for general demand response control has been considered to be about 30 min. Therefore, in general demand response control, real-time performance was not required for notification such as a demand suppression of electric power. For this reason, many demand response control systems have adopted a periodic pull readout method from the customer system to the utility's system.

A general communication sequence between a utility or aggregation server, which is a demand response control server in conventional demand response control, and BEMS, which is a demand response control client of customer premises, is shown in Fig. 3.12a. BEMS reads all data periodically because it does not know when

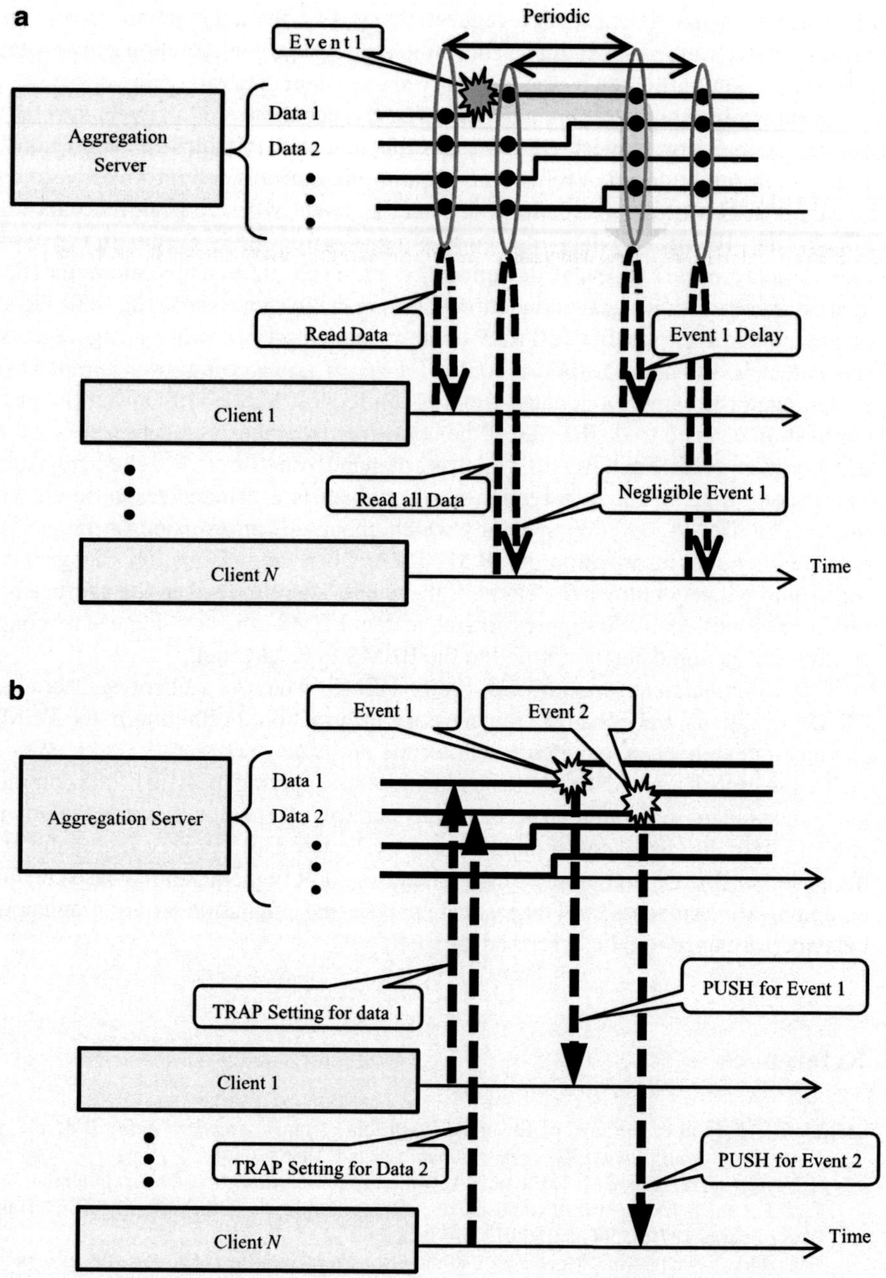

Fig. 3.12 Flexible and fast push notification of server event data by IEEE 1888 TRAP method

data such as demand suppression request changes on the aggregation server. The period was determined depending on the individual system configuration and network communication traffic, but it was generally about 1 min interval.

As shown in Fig. 3.12a, immediately after the occurrence of an event such as a demand suppression request, BEMS reads this in a state in which the aggregator's aggregation server does not reflect the event in the transmission information to the consumer's BEMS. If so, BEMS will detect an event when reading out data one minute later, BEMS will delay detection of aggregation server events. In the worst case, a delay occurs in event detection for one cycle time. This communication method is called "periodical pull readout". Figure 3.12b shows the IEEE 1888 TRAP communication service. In this TRAP communication service, when an aggregation server that is a demand response control server of an electric power company or an aggregator generates a demand suppression request for electric power, the push notification enabled to the BEMS of the consumer from the aggregation server that uses this as an event. First, BEMS, which is a demand response control client, registers TRAP conditions in the aggregation server, which is a demand response control server. The TRAP condition specifies which change of information on the server is to be the target information of TRAP. Then, when the information changes, the server immediately notifies the client of the event. Therefore, when the aggregation server generates a demand suppression request for BEMS subject to demand response control, this is immediately notified to the BEMS to be targeted.

This communication method can notify BEMS, which is a client registered as TRAP conditions, only changes in necessary information. Furthermore, the BEMS client can flexibly change the connection time of the network.

This section described the outline and features of applying the IEEE 1888 communication standard to communicate FastADR control information between the downstream part of the virtual power plant system, i.e. between the resource aggregator and the customer BEMS. A concrete and detailed FastADR implementation based on this communication standard, and theoretical analysis and simulation on communication delay performance will be described Chap. 8.

References

1. Technical report of the Inst. of Electrical Engineers of Japan, *Standard Analysis Model for Power Supply and Demand/Frequency Simulation*, No. 1386 (2016)
2. C. Ninagawa, H. Asaka, M. Takahama, Asymmetric constant-rate transfer function model for FastADR feedback control of a cluster of building multi-type air-conditoners. IEEJ Trans. Power Energy **137**(8), 566–572 (2017) (in Japanese)
3. K. Suzuki, C. Ninagawa, J. Morikawa, Estimation on averaging effect of large-scale aggregation of DR transient responses of building multi-type air-conditioning facilities. IEEJ Trans. Power Energy **138**(7), 582–590 (2018) (in Japanese)
4. OpenADR Alliance. https://www.openadr.org/
5. M. Piette et al., *Open Automated Demand Response Communications Specification (Version 1.0)*, LBNL Report 1779E (2009)

6. OASIS Standard, *Energy Interoperation Version 1.0, OASIS Committee Specification Draft 01, 2010*. http://docs.oasis-open.org/energyinterop/ei/v1.0/os/energy-interop-v1.0-os.html
7. IEC standard, *Reference Architecture for Electric Power Systems*, IEC 61850 (2011)
8. IEEE standard for Ubiquitous Green Community Control Network Protocol, IEEE1888-2011 (2011)
9. Open System for Energy Services (OS4ES). http://www.os4es.eu/
10. IEC Standard, *Specific Communication Service Mapping (SCSM)—Mappings to MMS (ISO 9506-1 and ISO 9506-2) and to ISO/IEC 8802-3*, IEC 61805-8-1 (2004)
11. ISO Standard, *Industrial Automation Systems—Manufacturing Message Specification*, ISO9506:2003 (2003)
12. IEC standard, *Communication Networks and Systems for Power Utility Automation-Part 6: Configuration Description Language for Communication in Electrical Substations Related to IEDs*, IEC 61850-6 (2009)
13. IEEE standard, *Ubiquitous Green Community Control Network Protocol*, IEEE1888-2011 (2011)
14. Green University of Tokyo Project. http://www.gutp.jp/
15. C. Ninagawa, H. Yoshida, S. Kondo, H. Otake, Data transmission of IEEE1888 communication for wide-area real-time smart grid applications, in *International Renewal and Sustainable Energy Conference IRSEC'13*, pp.1–6 (2013)
16. C. Ninagawa, T. Iwahara, K. Suzuki, Enhancement of OpenADR Communication for Flexible FastADR Aggregation Using TRAP Mechanism of IEEE1888 Protocol, in *IEEE International Conference on Industrial Technologies, ICIT 2015*, Seville, Spain (2015)
17. T. Iwahara, T. Yamada, C. Ninagawa, Smart grid OpenADR communication Realized by Using IEEE188TRAP Mechanism's Flexible Server-Push Capability, in *IEEJ Technical Meeting on Smart Facilities, SMF-15-014* (2015) (in Japanese)

Chapter 4
Components of Virtual Power Plant

Check for updates

4.1 Power Storage Battery

4.1.1 Battery Energy Storage System

Chapter 1 explained that the fluctuations in renewable energy generation is a major cause of the imbalance between supply and demand. By connecting the battery system as a power source to the grid, it is possible to reduce the fluctuations of the renewable energy generation. That is, if the power supply is insufficient, the battery can be discharged to reduce the power demand. And if the power supply is excessive, the battery can be charged to increase the power demand, thereby contributing to the balance between supply and demand [1].

Table 4.1 shows the various type of batteries that are currently used as energy storage device. Lithium ion batteries, sodium sulfur (NAS) batteries, and lead acid batteries are mostly used as energy storage devices. Characteristics are different of each type of battery. For selecting a battery suitable for the energy storage device of a virtual power plant, the characteristics to be considered are output, capacity, charge/discharge efficiency, durability (life), charge/discharge rate and cost.

The C-rate (Capacity Rate) R_C is the ratio of the discharge (charge) current value I_C [A] to the battery capacity C_B [Ah]. For example, 1.0 C-rate of a battery means, it has the ability to complete the charging in one hour.

$$R_C = \frac{C_B}{I_c}[\mathrm{C}] \tag{4.1}$$

Furthermore, in order to provide energy services using batteries, it is essential to understand each storage battery's present State of Charge (SOC) to adjust the supply and demand by each battery. The capacity of the battery declines due to repeated charge and discharge (cycle life) and long-term storage (calendar life). But the capacity that takes into account the decrease in capacity from the initial capacity is considered as usable battery capacity C_{USA} [Ah]. And SOC is the ratio of the

C. Ninagawa, *Virtual Power Plant System Integration Technology*, Power Systems,
https://doi.org/10.1007/978-981-16-6148-8_4

Table 4.1 Performance comparison of storage batteries used as power storage devices

		Lithium ion battery	Sodium sulfur battery	Lead acid battery
Scale	Power	6–40,000 kW	500–50,000 kW	3–5000 kW
	Capacity	48–40,000 kWh	3,600–300,000 kWh	10–10,000 kWh
Charging/discharging efficiency	Single battery	95%	90%	85%
	In system	86%	75%	75%
Durability	Life span	10 Years	15 Years	17 Years
	Life cycle	10,000–15,000 cycle	4500 cycle	4500 cycle
Weight energy density		92–585 Wh/kg)	87 Wh/kg (786 Wh/kg)	25 Wh/kg(167 Wh/kg)
Volume energy density		176–3350 Wh/L)	83 Wh/L(1000 Wh/L)	62Wh/L(720 Wh/L)
Charge rate		8C	0.13C	0.3C
Discharge rate		Same as above	0.17C	0.6C
Charge rate (performance)		0.5–5.0C	0.13C	0.13–0.30C
Discharge rate (performance)		Same as above	0.17C	0.13–0.43C

battery dischargeable amount C_{DC} [Ah] and usable battery capacity C_{USA} [Ah].

$$SOC = \frac{C_{DC}}{C_{USA}} \times 100[\%] \tag{4.2}$$

Here, assuming that the relationship between battery chargeable amount C_{CA}[Ah], dischargeable amount C_{DC} and usable capacity C_{USA} is as Eq. 4.3.

$$C_{USA} = C_{DA} + C_{CA}[\text{Ah}] \tag{4.3}$$

Figure 4.1 shows a conceptual diagram of the SOC of a battery. It is noted that the chargeable amount of the storage battery is between the 100% charge amount and the maximum charge amount of the usable battery capacity. And the dischargeable amount is between the 0% charge amount and the minimum charge amount. In any case, the demand adjustment power of the battery is charge rate and SOC between the maximum charge and the minimum charge.

For the storage batteries in Table 4.1, SOC cannot be measured easily. Measuring state of charge by voltage is simple, but it can be inaccurate because cell materials and temperature affect the voltage. The most blatant error of the voltage-based SOC occurs when disturbing a battery with a charge or discharge. The resulting agitation distorts the voltage and it no longer represents a correct SOC reference. To get

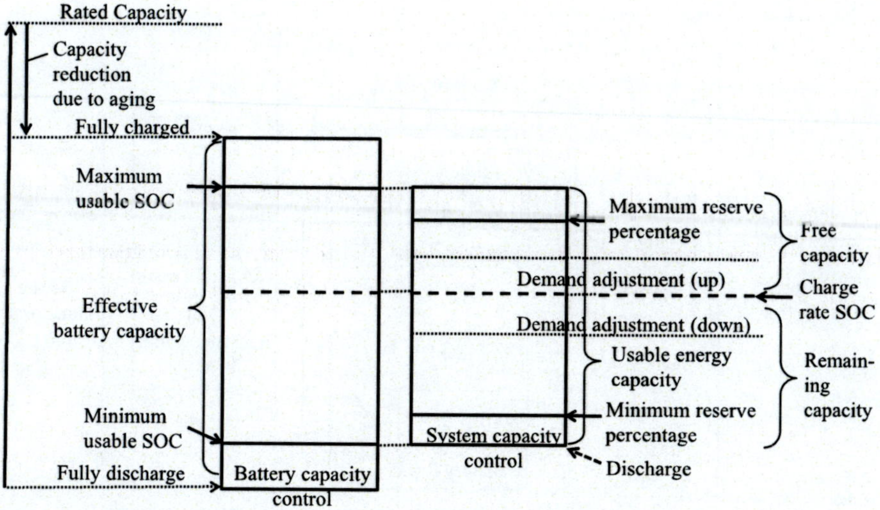

Fig. 4.1 Conceptual diagram of SOC of storage battery (lithium ion battery etc.) (The Institute of Electrical Engineers of Japan: JEC-TR-59002: Standard Specification for Energy Service with Battery System (2018), Partially Modified from p. 13)

accurate readings, the battery needs to rest in the open circuit state for at least four hours; battery manufacturers recommend 24 h for lead acid battery. When measuring SOC by open circuit voltage (OCV), the battery voltage must be "floating" with no load attached. This is not the case with modern power system.

Modern measuring units use coulomb counting to estimate SOC by measuring the in-and-out-flowing current. Ampere-second (As) is used for both charge and discharge. Coulomb counting works well, especially with Li-ion battery that offer high columbic efficiency and low self-discharge. This method also has some deficiencies, but still considerable.

4.1.2 Grid Connection of Battery System

Figure 4.2 shows an example of connection scheme of battery system installed in the facility of the customer. The battery system consists of a battery rack that stores the storage batteries, a Battery Management Unit (BMU), a Power Conditioning System (PCS), and a storage battery control. Power conditioning system performs DC-AC conversion. In addition, to prevent the damages the battery system is connected to the power grid via a circuit breaker. Furthermore, a step-up transformer is used according to the voltage level of power system.

As shown in Fig. 4.2, when the battery system is connected to the grid as a energy source and power flows to the grid side. In addition, as shown in Fig. 4.3, the power supply service provider receives electricity generated and procured by retail

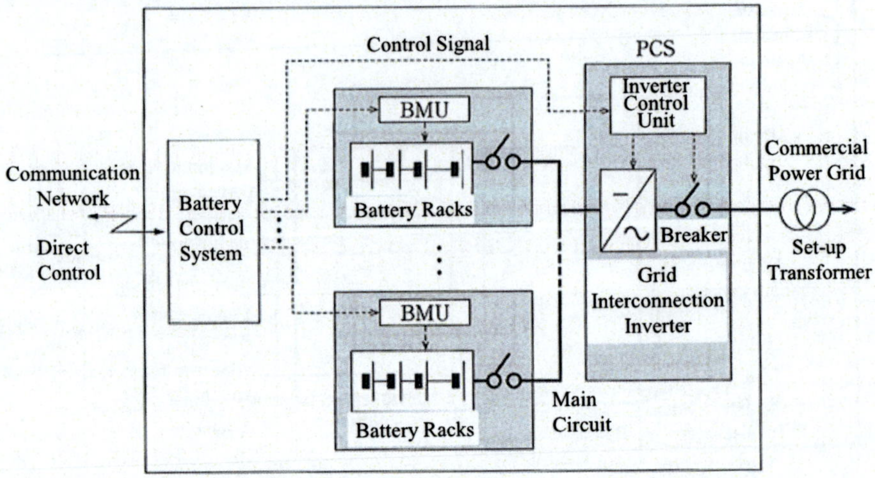

Fig. 4.2 General configuration of storage battery system (power storage device)

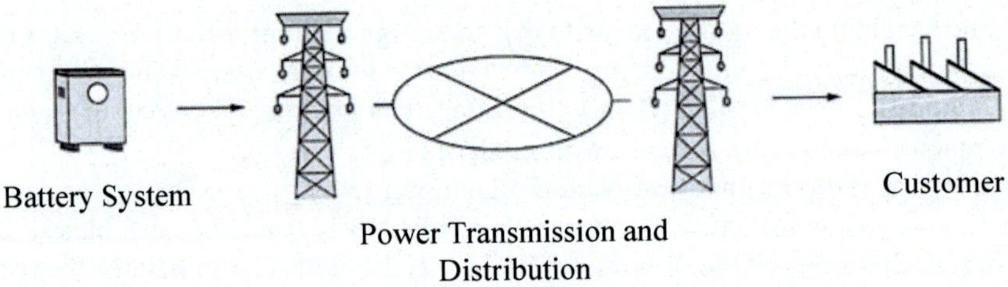

Fig. 4.3 Conceptual diagram of connection supply service

electric utilities etc. through the transmission and distribution network of the general transmission and distribution company. On the other hand, the power adjustment service. As shown in Fig. 4.4, the general transmission and distribution company receives electricity related to the supply service provider generated by the power generation company, etc. Generation company responsible to provide the amount of electricity offered by supply service provider including peak power that can overcome the demand fluctuations.

Supply demand adjustment power is the amount by which a general transmission and distribution company receives electricity for a specified wholesale supply from a demand suppression contractor, and simultaneously offers to the demand suppression contractor at the same time through the transmission and distribution network Supply of electricity. In addition, as a result of demand restraint, when the demand

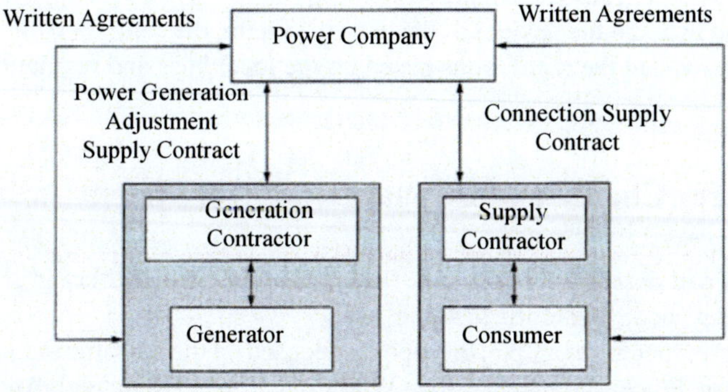

Fig. 4.4 Conceptual diagram of connection supply service

restraint amount runs short, it also includes replenishing the quantity of the shortage of electricity.

And in order to conclude the supply contract, the following requirements need to be fulfilled [x].

1. The electricity to be used for retail electricity business is the one related to the generation amount adjustment supply or the electricity to be provided for the temporary transmission supplied by the general transmission and distribution company.
2. The contractor can supply electricity according to the demand value of the demander.
3. The customer must electrically connect the electrical installation to our supply facility in compliance with the following matters: Technical standards set by law, other laws, etc. Technical requirements specified in the separate volume of the consignment Supply. Method that we consider to be technically appropriate, taking into consideration the conditions of our supply facilities, etc.
4. When supplying at high pressure or extra high pressure, the contractor and the customer should follow the power supply command from our company.
5. The contractor shall have the consumer comply with the provisions on the consumer in the terms of the consignment provision, etc.
6. Accept that the customer complies with the provisions on the customer in the consignment provision, etc.
7. If the customer confirms that the customer receives power supply from the Company or other contractors, the contractor is the customer required for the implementation of the connection supply without the Company notifying the contractor in advance. Use the information from the Company for the purpose of serving as a retail electric power company of the Company, or consent to the provision to the other contractor.

From the above, in order to connect the storage battery to the system, it is necessary to understand in detail the devices of the battery system, the contract with the electric power company, and the regulations based on the local laws and regulations.

4.2 Battery Charge and Discharge Control System

This section will describe the method of charge/discharge control of grid connected battery system according to the demand.

Taking as an specific example, this book is focused on the pre-time market of JEPX (Japan Electric Power eXchange: Japan Wholesale Power Exchange). And describes how to plan and sell the amount of power generation (including negawatts) generated by the virtual power plant system. This market has already been established, and plans to demand and supply adjustment by charge/discharge control of the battery system.

4.2.1 JEPX Hour-Ahead Market: Japan Case

The Japan Electric Power Exchange (JEPX) started wholesale power trades in 2005, and the hour-ahead market was opened in 2016. The current one-hour-ahead market can trade until one hour before the actual delivery of power. As of October 2018, the JEPX trade volume for one-hour-ahead market was 5.0%.

This one-hour-ahead market trading divides a day into 48 slots of every 30 min. Until one hour before the delivery time, trading is conducted according to the Zara trading method (trading method in which contracts are made one after another starting from one that meets the conditions of the selling order and the buying order) [2]. The trading volume unit is 0.1 MW (50 kWh in 30 min). As an example, Fig. 4.5 shows the actual trading time series of August 6, 2018.

Here, we introduce the rules [2] relating to the pre-time trading of JEPX. JEPX provides various types of markets bulletin board such as spot market, forward market, hour-ahead market, and bulletin board market. The spot trading market buys and sells electricity for 30 min for the next day, and the one-hour-ahead trading market buys and sells electricity for 30 min after 1 h.

Furthermore, JEPX's power trading must be a trading member. Also, all trades of JEPX are conducted through the computer system provided by JEPX. The computer system is provided with a Web version and a Web API version.

The traded power is delivered through the power network managed by the general transmission and distribution company. But it must be in accord with the relevant regulations defined by the OCCTO (Organization for Cross-regional Coordination of Transmission Operators, hereafter referred as; power transmission coordinator) and the contract provision etc. defined by the general transmission and distribution company. Furthermore, the connection supply contract, power generation adjustment supply contract or demand control contract must be registered to JEPX for each

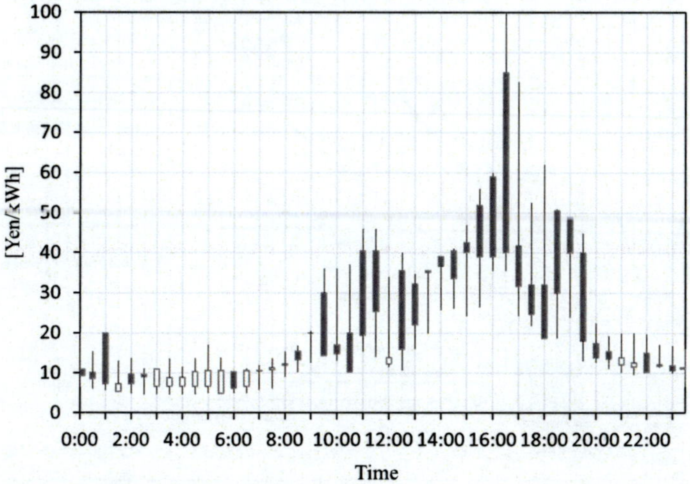

Fig. 4.5 Actual trading of the JEPX 1 h ago market (August 6, 2018) (Japan Wholesale Power Exchange website: created from market transaction results before fiscal year 2030)

control area defined by the power transmission coordinator. It is also necessary to deposit some charges which are determined by JPEX.

By the way, in one-hour-ahead trading, buying and selling of electricity for 30 min after 1 h, and delivering the electricity on delivery time are made according to the method defined by JEPX. The 30 min of delivery is called one product, and the transaction time for each product is 5 pm on the day before the product belongs, and the end time is 1 h before the delivery time zone. Every day bidding process of sale and purchase of the one-hour-ahead transaction is carried out for all day. The bid price and unit for the transaction are: 1 kWh for the asking price, 0.01 yen for the asking price, 50 kWh for the transaction unit, and 50 kWh for the delivery unit as in Fig. 4.6.

Contract by the field method every product. However, if trading bids in different areas become candidates for execution, in order to confirm to the power transmission coordinator whether or not the power transmission line between the areas related to the trading concerned, the following execution processing is temporarily stopped, and after confirmation, Bids will be executed only within the range where the answer is possible. After that, the paused processing is resumed immediately.

After this, JEPX notifies the execution result of the pre-time transaction (bid number, execution number, product, execution date/time, execution price, contract price and contract price). The trading member who receives the promised results notifies the power transmission coordinator of the sales and procurement plan from the promised results. Then, when the corresponding time comes, power is delivered through the power network according to the contents of the agreement. The fact that power has been delivered is confirmed via a smart meter (a digital power meter that can send data to a remote location). If the delivered power is less than the amount

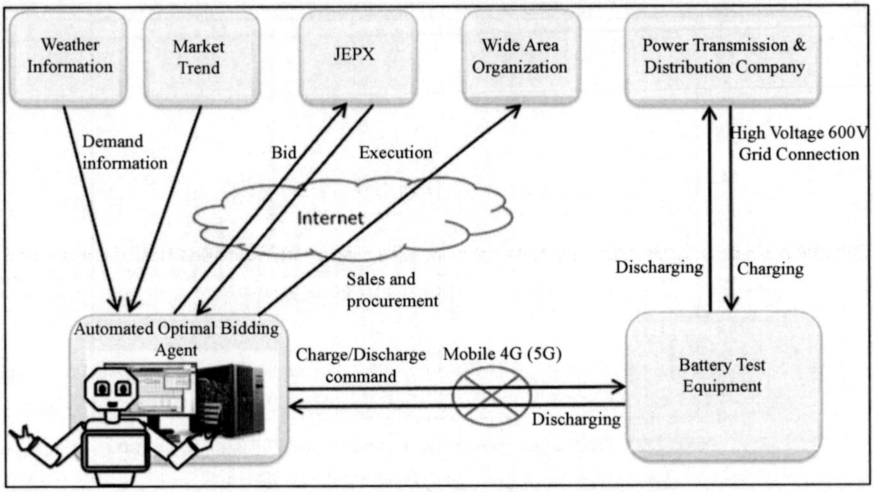

Fig. 4.6 Example of a storage battery system that performs power exchange with JEXP

of electricity promised and notified in the sales and procurement plan, it will be compensated by the general transmission and distribution operator, but a charge for the shortfall will be charged (this is called an imbalance fee).

Also, in the pre-hour transaction, the sales price (product of selling fixed amount and selling price), buying price (product of buying commercial fixed amount and selling price), and trading fee are targeted for settlement, every business day at 8 am the previous day. The contract between 12:00 pm and 12 pm will be closed, and the trading price and trading fee for each trading member etc. will be confirmed and notified. The settlement date is 2 business days after the start date of the notification date.

It goes without saying that in order to carry out such JEPX pre-time market transactions 24 h a day, 365 days a day, computer automation is essential, not human operations.

4.2.2 Automatic Optimal Bid Agent

Computer-automated system that buys and sells power generated by his own virtual power plant is called "automatic optimal bidding agent". This automatic optimal bidding agent predicts JEPX's price before the time market "power price forecast AI", AI group such as "sell and buy AI" to consider the SOC of the storage battery to judge buying and selling, "bidding to the time before market" "Procedure". Automatic control group performs "sales/procurement report" to power transmission coordinator, "discharge command to storage battery", IoT group performs "acquisition of status", and accounting group performs "clearing" of transaction value and fee.

Thus, delivering the power of the storage battery system via the power network of a general transmission carrier by forecasting price fluctuations in the market one hour-ahead of JEPX and buying and selling. It is a time zone where the demand for power is increasing. It is equivalent to supplying power to and storing power in a time zone with enough power.

In other words, if the power of large number of batteries are aggregated, it can be said that it is a virtual power plant that adjusts supply and demand (Fig. 4.7) by charging and discharging the batteries using the power trading system through the market.

Since the power demand predicted by the power company for a long time is very important for a stable supply of electric power. It has been carried out using the historical data and predicted data which is based on weather factors and social factors. Based on demand prediction, a power generation plan is scheduled that is a major feature of daily power supply and demand operation.

Therefore, JEPX's spot market calculates the deadline for trading of electricity to be delivered the next day at 10 o'clock every day. However, on the same day, sudden weather changes caused demand to exceed the previous day's expectations, the amount of power generated by solar and wind power plants did not increase, and the expected amount of power could not be procured. Failure to supply as planned may occur due to a malfunction of the power plant or transmission system.

For this purpose, there is a market one hour before in JEPX, and a mechanism is provided that can resupply the surplus and shortage of supply and demand that occurred one hour before that day. Therefore, the transition of the electricity price in

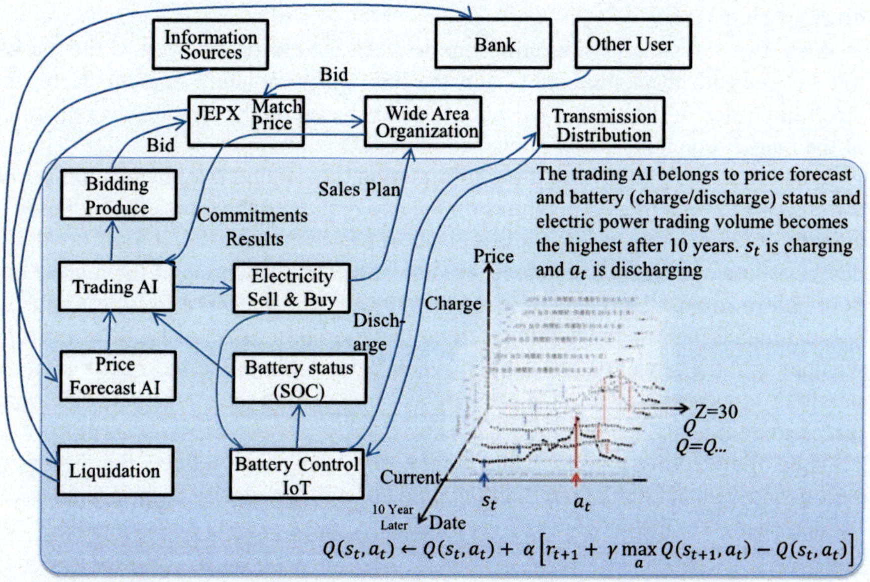

Fig. 4.7 JEPX automatic optimal bidding agent

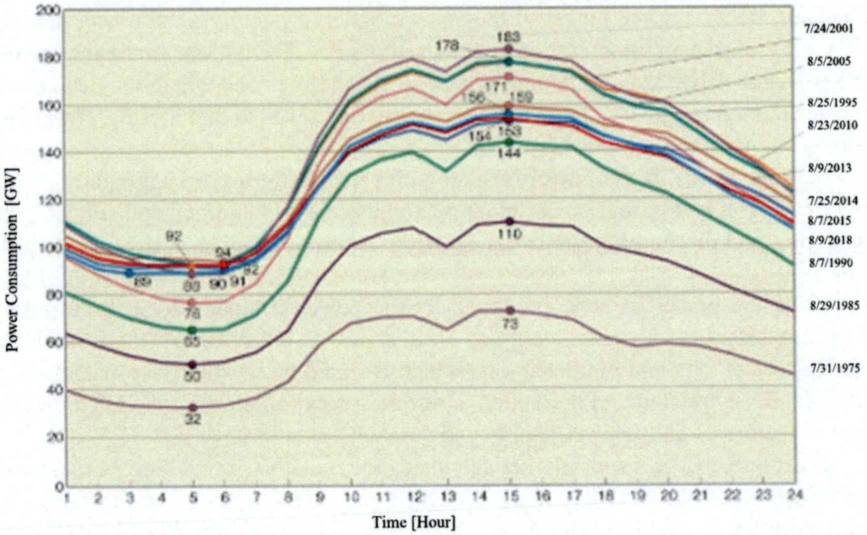

Fig. 4.8 Electric demand in summer days (Electric Industry Association: Transition of usage of electricity on the day of maximum power generation day Quote from www.fepc.or.jp/enterprise/jigyou/japan/)

the market one hour-ahead is different from the old daily curve electricity demand (Fig. 4.8), and the electricity in the time zone where the electricity demand is rising is not high. The power price in the time zone when the difference between the two diverges increases.

Also, Fig. 4.9 shows the relationship between the electricity price of the market one hour-ahead, the temperature, and the total solar radiation amount. Here, the electricity price of the market one hour-ahead is based on the national average price of the market one hour-ahead published by JEPX.

Looking at the graph from these information, it is found that there is almost no difference between the temperatures on the two days, and the total solar radiation is a gentle curve on August 1st, while it peaks at 13:00 on August 2. It turns out that it descends linearly. In such a weather change, the electricity price of the market one hour before August 2 has a peak once at 16:30 and then at 19:00 as it is on August 1. It is estimated that the temperature fell on the afternoon of August 2 due to cloudy weather, and the power demand in the evening was expected to be smaller than the previous day. It is thought that electricity was procured in the market and electricity prices soared.

Thus, the electricity price in the market one hour-ahead reflects the electricity supply and demand on the day, and the prediction of the electricity price due to the sudden change of the weather is a very useful technology.

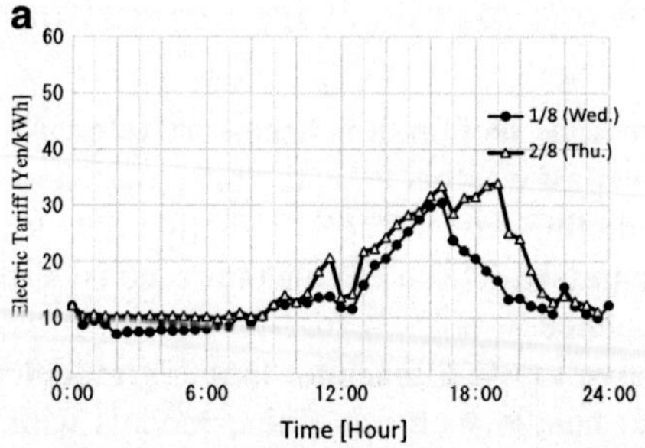

(a) JEPX one hour ahead of market August 1, 2018 (Wed) - August 2 (Thu) (The Japan Wholesale Power Exchange website: made from the market transaction results before the fiscal year 2030)

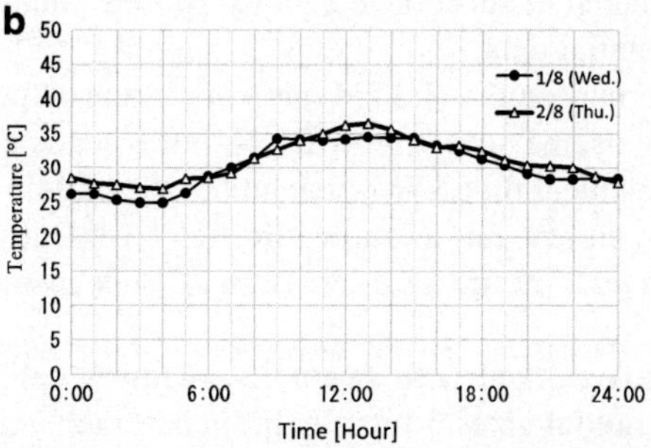

(b) Meteorological data (Tokyo): Temperature August 1, 2018 (Wed)-August 2, Thursday

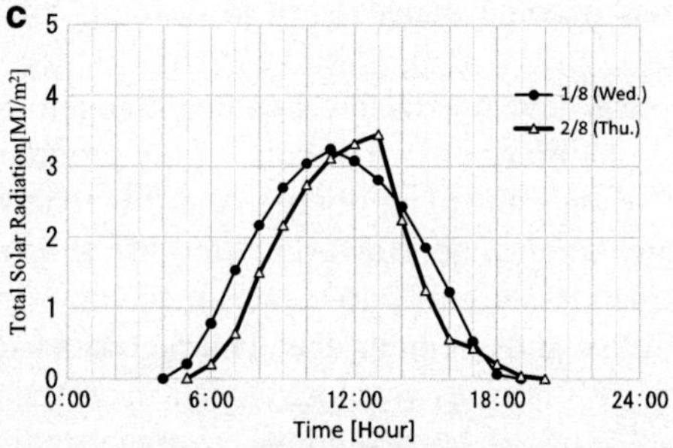

(c) Weather data (Tokyo): Total solar radiation August 1, 2018 (Wed) - August 2 (Thu) (Meteorological Agency website: Created from the hourly temperature of Tokyo on August 1, 2018, 2nd)

Fig. 4.9 The relationship between the electricity price in one-hour-ahead market, temperature and total solar radiation in the Tokyo area

4.2.3 Power Wholesale Market Price Forecasting Model

As mentioned in the previous section, it is necessary referring not only the previous market price but also the past weather data, the information of the economic activity scales simulated to predicting day and solar radiation data to predict the electricity market price more accurately. These elements are considered to time series data depend on the time sequence.

However, it is effective to use a machine learning model such as deep learning because it is difficult to handle such non-linear models with simple mathematical models.

A Recursive Neural Network (RNN) is one of the modeling technique that can capture, store, and make output such as the time series of the sequence data. A time series is a sequence taken at successive equally spaced points in time and it is not the only case of sequential data.

The Long-short Term Memory (LSTM) network is one of the most popular RNNs. LSTM is basically the same structure as RNN, but it leads a different function to calculate the hidden element. In LSTM each unit in the middle layer is replaced with an element called a memory unit as shown in Fig. 4.10. But there is no significant change in the I/O layer. The internal structure of this memory unit is shown in Fig. 4.11.

The memory unit is worked with five units, memory cells, inputs, input gates, forgetting gates, and output gates. Mutually influence each unit to each other, stores past inputs, forgetting, appropriately deal with time series data by each other output.

Recently, such a machine learning development environment has become widespread, and it has become unnecessary to develop a learning program from scratch.

At here, an example is developed using Chainer, an open source software library. Chainer is provided as a library of Python, one of the programming languages, and it is possible to use Python's abundant library. And it is relatively intuitive and easy to use GPU-based calculation code. However, after installing multiple libraries such as UDA outside Chainer, it is possible to create an actual program.

Weather data for Tokyo and one-hour-ahead market data from August 1st, 2018 to August 20th, 2018 of JEPX is used as evaluation data in presented example. The input variables are 4 variables of the frame number, average trading price, accumulated solar radiation and temperature in the one-hour-ahead market. Weather data and one-hour-ahead JEPX market data is collected from Japan Meteorological Agency website [3], and trading data [4] published to JEPX website respectively.

Figure 4.12 shows the prediction results. The bold line is the fluctuation of the average trading price for every 30 min frame in the actual one-hour-ahead market on August 22. And the black circle is the time when prediction was made by LSTM. For convenience of display, black dots are plotted for every 1 frame = 1 h. Thin lines and crosses represent predicted values 2–4 frames ahead by LSTM at the time of the black circle where the lines are connected. In other words, it is a graph that predicts

Fig. 4.10 Example of LSTM neural network model

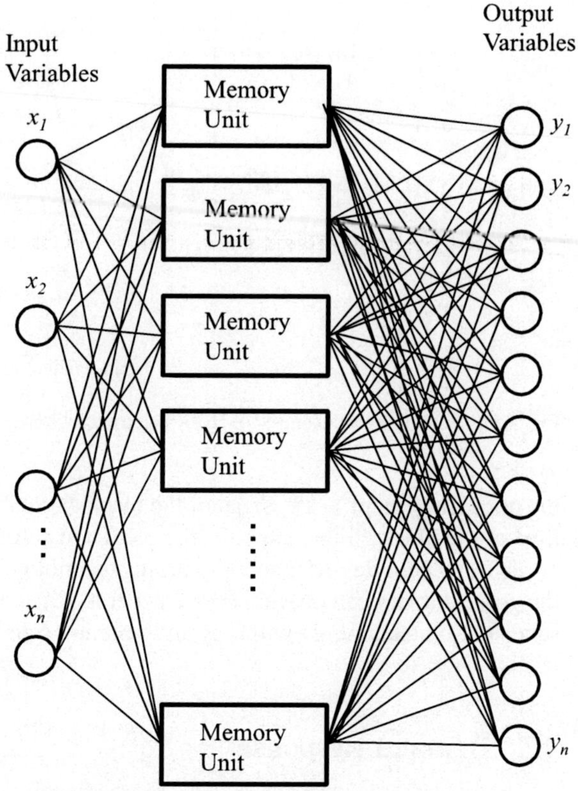

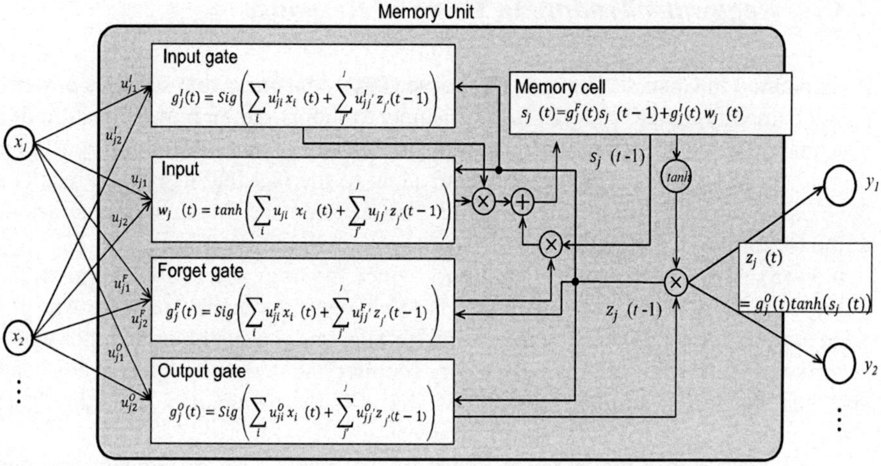

Fig. 4.11 Internal structure of memory unit

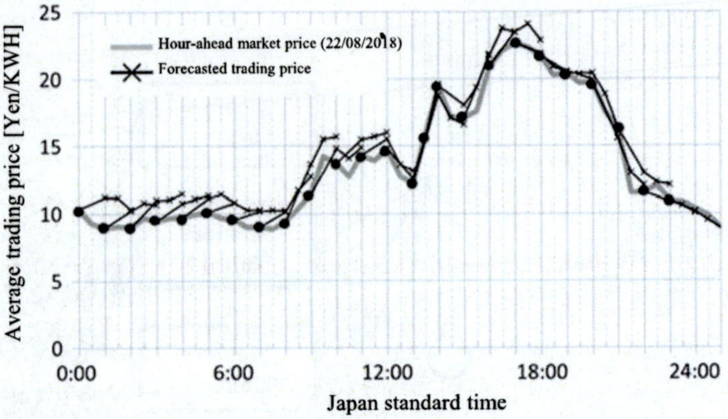

Fig. 4.12 Price forecast example of JEPX electricity wholesale market (one hour before the day)

the price movement 1–2 h ahead of the black circle from the data for 24 h before the time of the black circle, and plots the predicted value every hour.

Thus, by applying AI machine learning technology such as LSTM neural network, the power battery can provide power generation (including negawatts) at complying schedule. AI technology will play an important role in virtual power plant.

4.3 Demand Response

4.3.1 Negawatts Trading in Demand Response

As explained in Chap. 1, Demand Response (DR), a business that supplies power (a power company) makes a request to a consumer to suppress their power consumption. The amount of power reduction at this time is called negawatts. Although consumers do not actually generate power, it is equivalent to the fact that negawatts results in the balance of power supply and demand. The customers at this time also contribute to the formation of the virtual power plant.

Power saving and power saving are to reduce the total power consumption 24 h a day, 365 days, but demand response is to change the demand pattern. And it is done as the trading between the electricity supplier and the customer. In other words, according to "the request of the electricity supplier", the customer creates negawatts and "the electricity supplier buys it". So that it is the mechanism that can clearly obtain financial incentives for the customer.

Based on this demand response, there are two major types of demand response of virtual power plant. One is electricity demand type demand response (Fig. 4.13) and the other is incentive type demand response (Fig. 4.14).

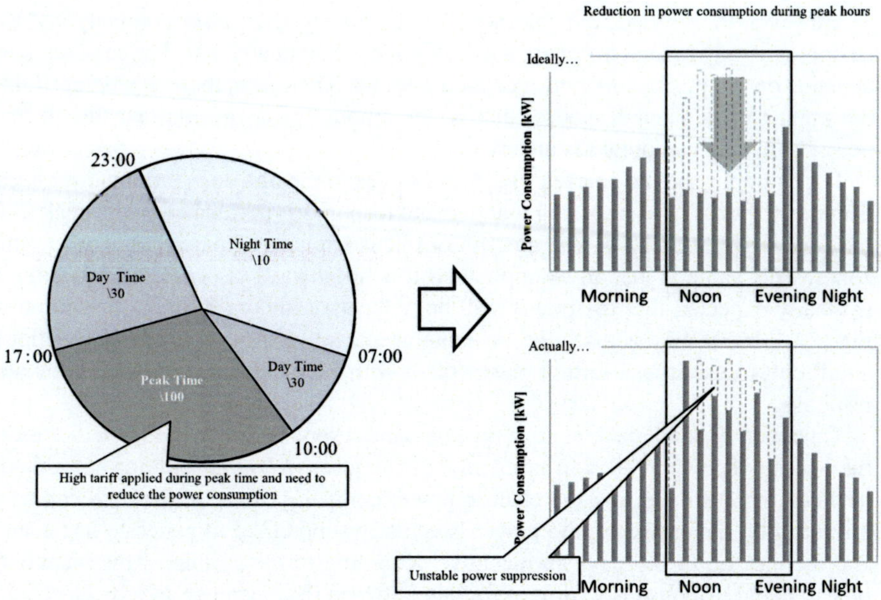

Fig. 4.13 Generation of negawatts by conventional electricity price demand response

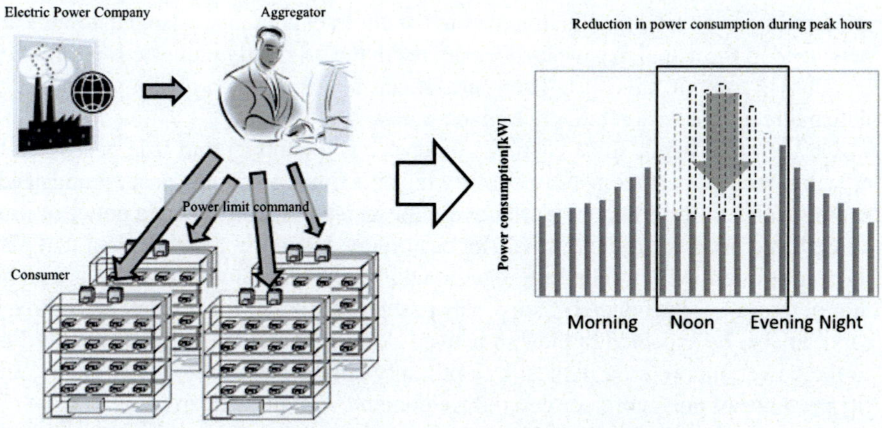

Fig. 4.14 Negative power generation by incentive type demand response

As shown in Fig. 4.13, the electricity supplier can increase the electricity charges for the time zone in which the power consumption is to be reduced. So that the power consumption reduction in that time zone with demand response can be reduced. This system is a relatively easy system for electric power companies, and it can apply demand response regardless of the number of consumers.

However, in this case, the contract between the electric power company and the customer is only related to "change in unit price". And how much change in demand depends on the positive response of the customer. Therefore, there is a demerit that the impact of demand response, that is, the amount of negawatts that the electric power company can obtain is uncertain.

For example, the pie chart in Fig. 4.13 is an example of a conventional long-term demand response, and assuming that demand response is performed in the peak time zone from 10:00 to 17:0. The electricity bill is tripled compared with other time zones. This value is just an example, and it is not always the case. At this time, it is ideally expected that the power will be reduced as shown in the upper graph of Fig. 4.13, but it may actually be unstable as shown in the lower graph. So that it is difficult to use it as a virtual power plant with such slow and uncertain demand response.

On the other hand, the incentive type demand response shown in Fig. 4.14 defines the amount of power reduction based on the contract as well as the bidding beforehand between the customer and the electric power company. Then, at the date and time promised by the customer, the power is actually suppressed to generate negawatts, and the electric utility pays an incentive according to the amount. This incentive-type demand response is a form of trading between the customer and the electricity supplier, in which the electricity supplier purchases the negawatts created by the customer's power reduction.

In the case of incentive type demand response, not only the incentives but also the amount of power reduction is included in the contract contents. Also, customers are penalized, if the actual negawatts does not reach the originally requested suppression amount (if commit fails) [5]. Therefore, it can be expected to obtain an impact of demand response with relatively high certainty.

On the other hand, since each customer needs a contract and it is necessary to calculate the compensation individually which is a time consuming process and needs some special skills to handle it effectively. In such a case, the balance between time and effort creates the issue to small-lot customers. However, a mechanism that play the role as intermediary called aggregator, which bundles many customers, as a mediator of trading, is introduced to solve this problem for small-lot customers. Aggregators can also be expected to play an active role to generate the negawatts according to the power suppression request by optimally combining multiple customers, and the presence of an aggregator can reduce demand with higher certainty.

4.3.2 Current Status of Negawatt Trading: Japan Case

Taking Japan as an example, there has been an effort similar to the electric tariff type demand response described in the previous section as well as in [4.8]. One of them is supply and demand adjustment contract. This is a mechanism for discounting the electricity charges on the condition that contract between the electric power company and the large demander meets the demand reduction when the supply and demand is tight. The other is Time of Use (TOU), which has a mechanism to create a charge menu that reduces nighttime electricity rates where the power usage rate is low and raises the daytime electricity rates that have a high utilization rate.

On the other hand, electricity demand type demand response is a more flexible version of these mechanisms. So, it can be applied without problems in present rules related to electricity business. While some retail electric utilities already offer a menu and negawatts trading in incentive-type demand response has been launched as a power source in 2018, but the formulation of rules in the system design of power system reform and environmental maintenance is needed. Currently, discussions are underway for appropriate system design [5].

At present, there are two broad categories of negawatts trading, which are institutionally designed in Japan (Fig. 4.15). One is that a retail electric utility procures from a customer (type 1①) or one that is procured from the same trader (type 1 ②), and the other is that a system operator procures from a retail electricity supplier/consumer.

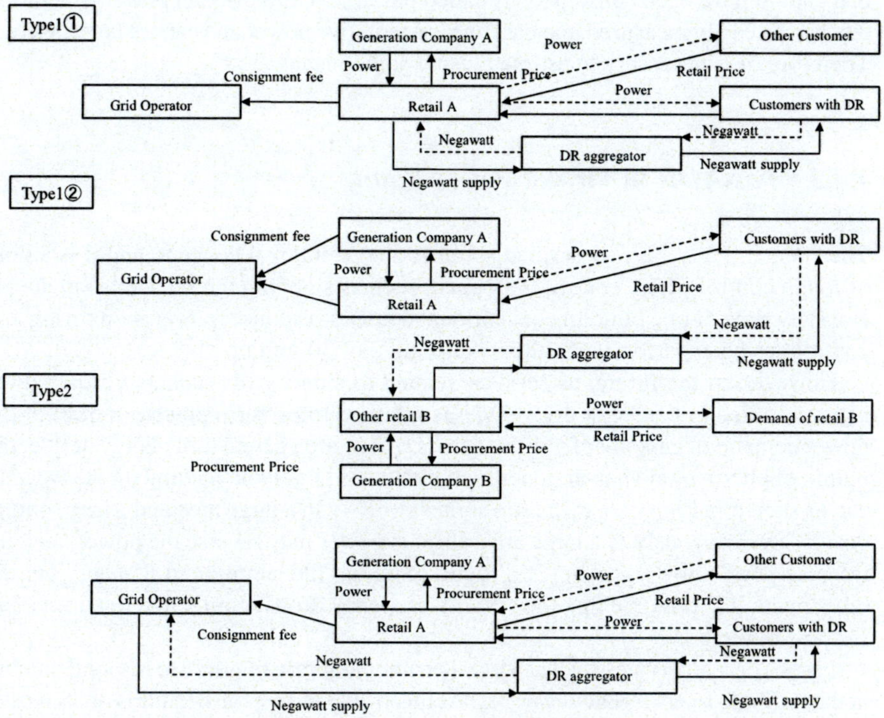

Fig. 4.15 Types of system design for negawatts trading (Comprehensive Resource and Energy Study Group Basic Policy Subcommittee, Power System Reform Subcommittee: Next-generation Transmission and Distribution Network Study Group: Material submitted by the 9th System Design Working Group Secretariat-Utilization of Negawatt Trading (2016), p. 3 (Partially modified)

Type 1① in Fig. 4.15 is a model in which retailer A issues a demand response request to the consumer. And the consumer procures the Negawatts actually generated by the consumer through an aggregator and commits with the system operator (between retailer and consumer). On the other hand, Type 1② is a model (between retail and retail) in which retailer B commits by procuring the Negawatts created by customers contracting with retailer A through the aggregator. Type 2 is a model in which a system operator issues a demand response request directly to a customer, procures it via an aggregator, and uses the obtained Negawatts as a demand–supply adjustment power.

By the way, in negawatts trading, it is necessary to pay the customer incentives according to the generated negawatts. So, it is necessary to calculate the amount of power reduction accurately. Here, the amount of power reduction refers to the difference between "the amount of power that should have been consumed if the consumer did not receive the power suppression due to demand response" and the actual amount of power consumption (the amount of actual demand). This "power that should have been consumed" is called baseline. Unlike actual power generation that can be easily measured, the baseline is a tentative power and cannot be measured. Therefore, it is necessary to estimate using some means.

4.3.3 FastADR in Virtual Power Plant

The "demand response" described so far is a conventional concept, and it is a kind of rough control where control is stopped, such as stopping the selected load device in a slow time of one hour. In conventional demand response, power monitoring and control has been performed with response time of 30 min or 1 h.

However, in the future, as services related to smart grids such as virtual power plants handled in this book diversify, it is required to perform precise control of the power demand of customer facilities across a wide area in real time. For example, the output of photovoltaic power generation fluctuates largely depending on the weather etc. as shown in Fig. 4.16. If the sun shines strongly in a large area and a large output occurs simultaneously in a large area, surplus power may reverse the power network (reverse power flow), making it difficult to adjust the supply and demand. On the other hand, it is possible that the system frequency drops rapidly due to an accident in some of the power grids.

In such cases, services such as virtual power plants are required to respond quickly and contribute as emergency reserves. At this time, the response in units of 30 min or 1 h, such as the conventional demand response, cannot be made in time. In addition to the response speed, the optimal control of the customer's facilities over a wide area can be obtained simultaneously.

Therefore, as shown in the conceptual diagram in Fig. 4.17, the automatic demand response (Automated Demand Response: ADR) performs monitoring and control in real time by automatically aggregating information such as forecast and results of power demand of customers via a computer network. ADR is a control method that

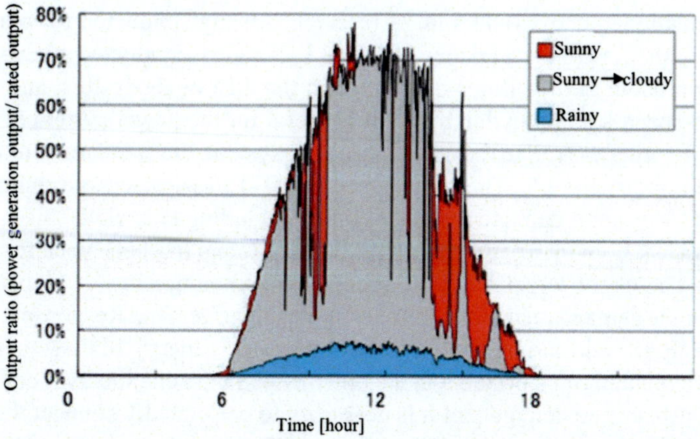

Fig. 4.16 Output fluctuation of photovoltaic power generation (Next Generation Transmission and Distribution Network Study Group: Towards the Construction of the Next Generation Transmission and Distribution Network to Realize a Low-Carbon Society (2011), Partial Modification from p. 12)

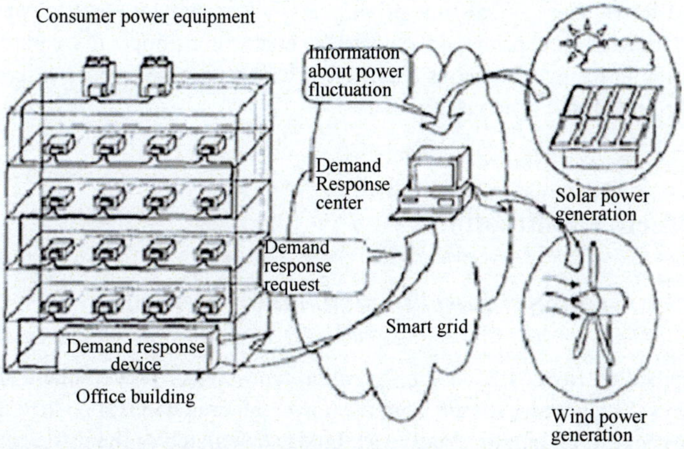

Fig. 4.17 Output fluctuation of photovoltaic power generation (Next Generation Transmission and Distribution Network Study Group: Towards the Construction of the Next Generation Transmission and Distribution Network to Realize a Low-Carbon Society (2011), Partial Modification from p. 12)

assumes that the demand response center on the grid side and the demand response device on the demand side perform direct communication control via the network. There is no need for a human administrator, and in consideration of the operation of a large number of remote users, this is realized on a so-called M2M (Machine to Machine) system in which computers communicate with each other bi-directionally.

In particular, ADR control that responds in minutes, sometimes in tens of seconds, is called FastADR. FastADR is an ADR that can function as an emergency reserve just as mentioned above, and a lot of research is being conducted for practical use. In the US, the Lawrence Berkeley National Laboratory conducted an experiment to shut down about 200 facilities distributed in the area in about 20 s, and confirmed that high-speed load reduction by ADR is useful for the power system [6].

Now, in order to realize the above FastADR system, the standard communication standard is indispensable. Therefore, OpenADR [7] was devised as a communication standard for general demand response control including FastADR.

OpenADR is an international standard of ADR, and has been used for more than 7 years. The client/server form of OpenADR communication is adopted, and the server for the demand response control of aggregator (Demand Response Automation Server: DRAS) and the client on the customer side (Energy Management System: EMS) are regularly performed. On the other hand, in the client/server configuration, there is also a disadvantage that it is necessary to periodically connect the customer response device to the network. Therefore, notify from the server. FastADR control using Push method with web socket is also considered only when communication is required [8]. Details of these OpenADR and concrete construction techniques are described in Chap. 8.

Also, as a system that enables consumers to utilize high-speed communication as described above, there is real-time pricing (RTP) that changes electricity rates in real time [9, 10]. Real-time pricing is expected to enable consumers to operate economic power suitable for instantaneous electricity charges. Please refer to references [11–13] for details of real time pricing.

4.4 Baseline Estimation

4.4.1 Conventional Baseline Estimation Method

As described in Sect. 4.2, in the incentive-type virtual power plant, the electric utility pays the customer for the negative power generated in response to the demand response request. Here, the negative wattage is calculated as the difference between the amounts of power consumption when not suppressed and when suppressed. This "power consumption without suppression" is called a baseline (Customer Baseline Load) [14].

Figure 4.18 shows an image of the amount of power suppression as an example, and the actual load is shown by a solid line and the baseline is shown by a dotted line. In this Fig. 4.18, a demand response is performed from 12:30 to 17:30, and the suppression of power consumption. If there is no demand response power reduction during this time zone, the baseline is "how was the power originally supposed to be consumed"? That is, the portion surrounded by the solid line and the dotted line is the negative watt, and the incentive type reward is paid according to the area of the negative watt (precisely, the accumulated amount in time slot units).

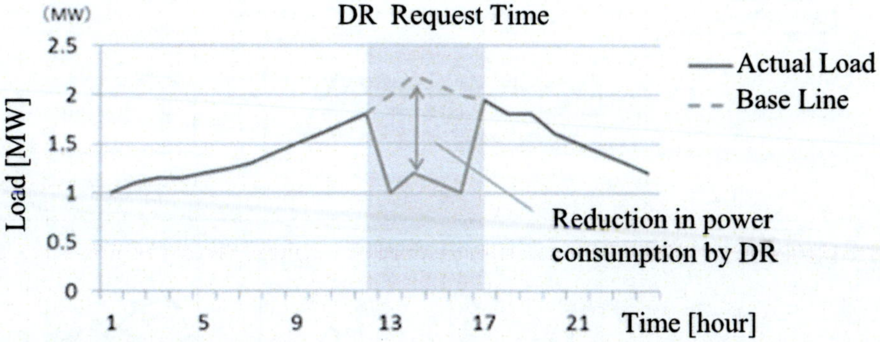

Fig. 4.18 Output fluctuation of photovoltaic power generation (Next Generation Transmission and Distribution Network Study Group: Towards the Construction of the Next Generation Transmission and Distribution Network to Realize a Low-Carbon Society (2011), Partial Modification from p. 12)

The problem here is that the calculation of negawatts depends on the baseline. If the baseline is underestimated, the calculated amount of negawatts will decrease, and the reward will also decrease, leading to a decrease in the profit of customers. Conversely, if the baseline estimate is high, the compensation will be high, and the burden on the electric utility will increase. Therefore, it is very important to estimate the baseline accurately.

However, data acquisition of baseline and actual load is alternative, and baseline cannot be acquired as data when power reduction occurs by demand response. Therefore, it is necessary to estimate using various data in the vicinity.

In fact, it is difficult to estimate correctly. The reason is that power consumption is time series data that has some relationship with the past history, and it is very difficult to capture the uncertainty of time series data [15]. It can theoretically be concluded that "it would have been" precisely, but it is difficult to convince. Baseline estimation is also difficult because it is a solution to the question, "If demand response request occurred two hours ago and power was suppressed, but what was the power consumption for two hours after that?".

Therefore, research to estimate baseline using statistical methods has been in progress since the 2000s. In these studies, it is basic to collect data when there is a demand response and when there is no demand response, and to process data in a fixed time width (time slot) as time series data summarized in one point. Conventionally, this time slot is normally 30 min to 1 h.

Thus, the baseline estimation methods conventionally proposed conventionally are roughly classified into three methods [16]. An averaging method, a regression method, a pattern matching method, and a data mining method (Fig. 4.19). Among them, the averaging method is often used, and it is easy to understand and implement using a simple method, such as averaging the power consumption at the same time in the last few days.

The regression method is a method of modeling power consumption with some formula using ambient temperature, room temperature, solar radiation, etc. Moreover,

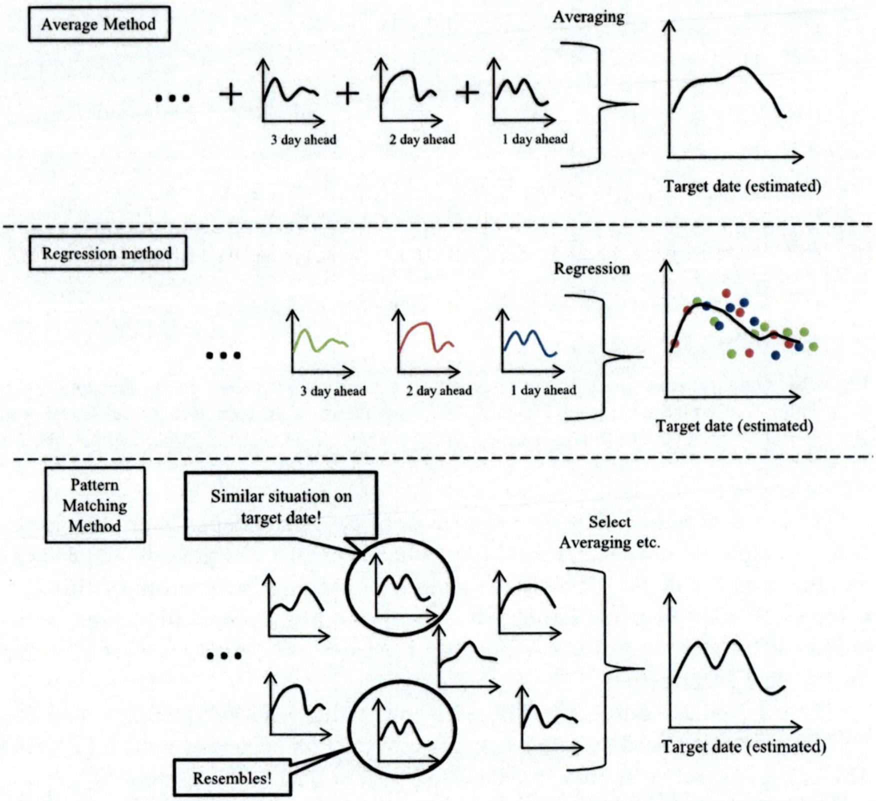

Fig. 4.19 Output fluctuation of photovoltaic power generation (Next Generation Transmission and Distribution Network Study Group: Towards the Construction of the Next Generation Transmission and Distribution Network to Realize a Low-Carbon Society (2011), Partial Modification from p. 12)

the pattern matching method is a method to find out and use the data when the time before and after the demand response occurs, the temperature, etc. are similar from the past data, and there is no demand response.

The conventional demand response is based on the assumption that the power control has a relatively long duration of 30 min to several hours, and a simple time frame averaging method is often used. The average method is a method of calculating the average value as the name for power consumption data at the same time several days before the demand response, and using it as a baseline estimation site. There are various ways to calculate the target, and the X of Y method [17], which averages X days before the Y days, and the moving average method, which averages Y days of data without demand response [18] and so on.

(A) X of Y method

The X of Y method is a method of preparing data for Y days without demand response going back from the day of demand response, and calculating a baseline from data

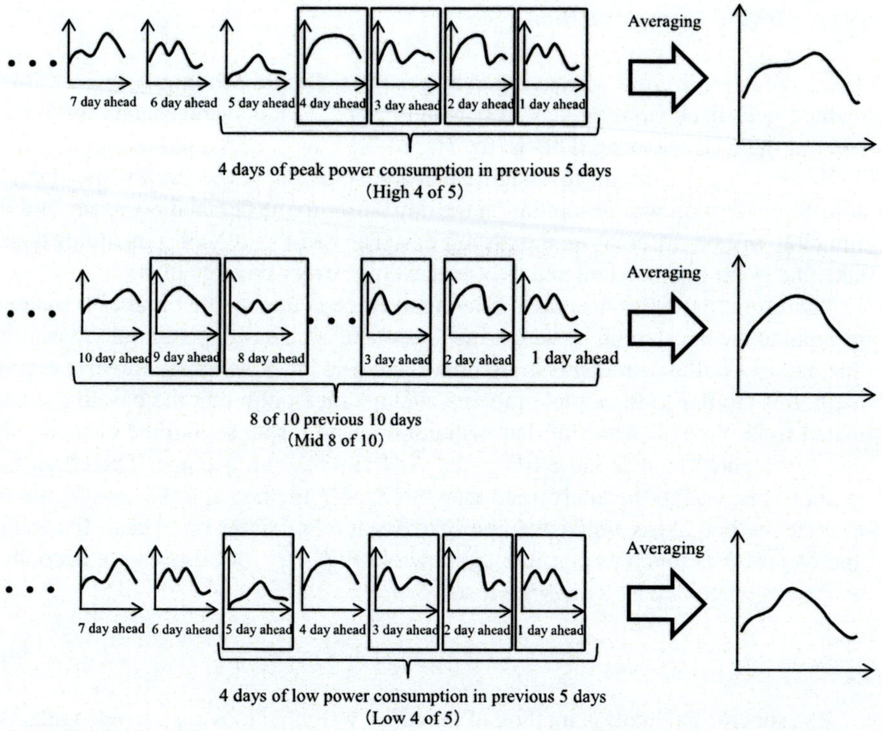

Fig. 4.20 Conceptual diagram of High X of Y, Mid X of Y, Low X of Y

of X days that satisfy the conditions. There are three conditions to select this X day as shown in Fig. 4.20.

One is High X of Y, which is the data of Y days, which is selected in descending order of total power consumption per day, and the average value of X days is used as a baseline. PJM in the US adopts High 4 of 5 on weekdays, High 2 of 3 on holidays, High 5 of 10 for High ISO, High 2 of 3 for CAISO, High 10 of 10 and High 4 of 4 for CAISO [17].

The second is Low X of Y [15], which was devised after that, and the total value of the power consumption per day is selected in ascending order of X days contrary to High X of Y, and the average value of X days is baseline It is assumed that Low X of Y is more accurate than High X of Y, but has negative "deflection" instead. In Low X of Y, three patterns of 4 of 5, 5 of 10 and 10 of 20 are used for X and Y.

The final third is Mid X of Y, where the total value of power consumption from day Y to day 1 with high values and low values are excluded by the same number, with the average value of remaining X days as the baseline is there. That is, (YX) mod 2 = 0, and Z = (YX)/2, data of X days selected is from data for Y days to total amount of power consumption in descending order of Z days, in ascending order. Mid 8 of 10, etc., are being studied by the US grid operator ERCOT etc. [18].

(B) Moving average method

There are several types of moving averages. The Simple Moving Average (SMA) method goes back from the day of demand response, and averages data for Y days without demand response. That is, the High/Mid/Low Y of Y method.

However, it is difficult to imagine that the situation of the supply and demand adjustment system will be similar on the day closer to the demand response, and the situation will be different on the distant day. The SMA and X of Y methods do not take that point into account and treat and average every day equally.

Therefore, the index-weighted moving average is the one that makes it easier to respond to the latest trend by weighting the data of each day exponentially, assuming that the data of the near day is more important. The index weighted moving average method is similar to the simple moving average method in that the baseline is estimated from Y days' worth of data without demand response, but the data for each day is weighted as it is more important Add and take an average. Therefore, it is easier to respond to the latest trend than the X of Y method and the simple moving average method. Also, unlike weighting in linear (continuous time) case, the weight approaches 0 as much as possible as shown in Fig. 4.21, but it does not become 0, so the influence of past data is not lost.

$$\tilde{w}_\tau = \lambda \cdot \tilde{w}_{\tau-1} + (1-\lambda) \cdot W_\tau \tag{4.4}$$

The specific calculation method of the index weighted moving average method is explained below. Letting τ be such that $0 < \tau \leq Y$ be a day, and the total load on that day be W_τ, the exponential weighted moving average load $\tilde{w}_\tau$ is expressed as: Here λ is a smoothing coefficient, and $0 < \lambda < 1$. As a result, the coefficient of the term $\tilde{w}_{\tau-1}$ is raised to the power of i. After defining the initial load $\tilde{w}_\tau$, this calculation is sequentially calculated from $\tau = 1$ to $\tau =$ Y, and the final value of $\tilde{w}_Y$ is taken as an estimate of the baseline. The smoothing factor and the initial load need to be determined by trial and error. Note that $\lambda = 0.5$, $\tilde{w}_O = 1/Y \cdot \sum_{j=1}^{Y} W_j$ in the New England Region system administration organization ISO-NE [15].

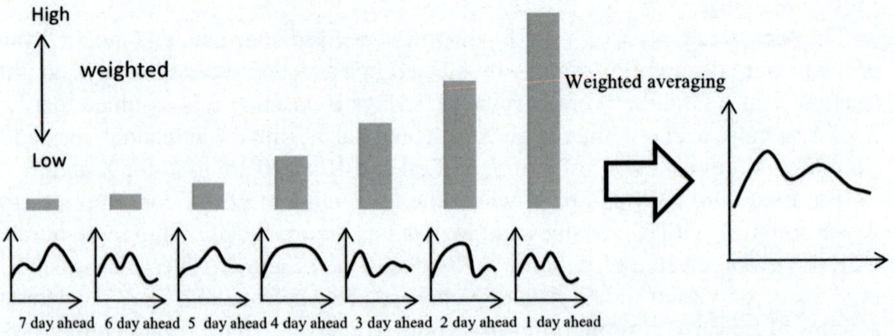

Fig. 4.21 Conceptual diagram of baseline estimation using exponential weighted moving average

In addition, there are accuracy, deflection, and fluctuation as a measure to measure the accuracy of the baseline estimation method [14–17]. Accuracy is expressed as the mean absolute error (MAE) or the relative root mean square error (RRMSE) between the baseline estimate and the true value. That is, in a time slot range s_s to s_e in which there is a demand response, let b_s be the predicted value of the baseline calculated in a certain time slot s, and b_s^* be the "true" baseline in that time slot. The MAE is as Eq. 4.5.

$$MAE = \frac{1}{s_e - s_s} \sum_{s=s_s}^{s_e} \left| b_s - b_s^* \right| \tag{4.5}$$

Also, assuming that the true baseline average value in the time slot range s_s to s_e of b_s^* is $\overline{b^*}$, RRMSE is as Eq. 4.6.

$$RRMSE = \frac{\sqrt{\frac{1}{s_e - s_s} \cdot \sum_{s=s_s}^{s_e} \left(b_s - b_s^*\right)^2}}{\overline{b^*}} \tag{4.6}$$

The larger the value, the larger the error of the estimated value. Next, deflection is a measure that represents the bias of the variance of the error.

$$Bias = \frac{1}{s_e - s_s} \sum_{s=s_s}^{s_e} \left(b_s - b_s^*\right) \tag{4.7}$$

When this deflection is large in the positive direction, it becomes easy to estimate the baseline largely, and conversely, when it is large in the negative direction, it becomes easy to estimate the baseline small.

In addition, the fluctuation is a measure to evaluate the robustness of the baseline estimation method, which represents the magnitude of the dispersion in the estimated value under various conditions. Specifically, the relative error ratio (RER) is calculated collectively for all time zones in which demand response has occurred? That is, Here, D is the total number of demand responses in the evaluation period, the subscript d is a reference number of the demand response request, and the variable with d as a subscript is the value at the time of the demand response. In other words, the baseline estimation accuracy will differ depending on the state of demand response as the RER is larger.

$$RER = \frac{1}{D} \sum_{d=0}^{D} \frac{\sqrt{\frac{1}{s_e^d - s_s^d} \cdot \sum_{s=s_s^d}^{s_s^d} \left(b_s^d - b_s^{*d}\right)^2}}{\overline{b^{*d}}} \tag{4.8}$$

4.4.2 Baseline Estimation of FastADR

In the conventional demand response, as mentioned earlier, the time slot has been taken as 30 min to several hours. It possible to implement the baseline estimation method in a form close to optimal without effected by the effects of fluctuation in power using averaging effect of long term data. However, in order to realize real-time power control for several minutes, FastADR control must be performed in several minutes. In the case of several minutes, the baseline estimation by the conventional averaging method is greatly affected by the rapid fluctuation. The estimation accuracy is significantly reduced, and it becomes impractical. For example, suppose you want to know the number of guests staying in a certain time slot between 12:00 and 13:00 in a certain store. If it is possible to calculate that a total of approximately 10 people stay in the same time slot in the past. It can be estimated that 10 people similar to the previous data will stay if viewed in 1 h slot.

However, when viewed in minutes, there is a large variation from day to day. Then, it turns out that it is not appropriate to simply use data in the past same time slot in minutes.

In recent years, various methods have been considered to estimate the baseline considering the past and present conditions. That concept fall into the regression method and the pattern matching method.

The regression method models power consumption with linear/nonlinear equations with multiple variables. The specific contents are explained in Fig. 4.22. The upper left of Fig. 4.22 is a scatter plot of certain x and y. Here, y can be represented by $f(x)$ by drawing a line as shown in the upper right of Fig. 4.22. A straight line is drawn on a two-dimensional plane, but it is also possible to draw hyperplanes in multiple dimensions as $\boldsymbol{x}$ $(\boldsymbol{x}_1, \ldots\ \boldsymbol{x}_{\mathrm{n}})$. That is, in the regression method, the power consumption is y, a plurality of variables are vectors, and a function f such that

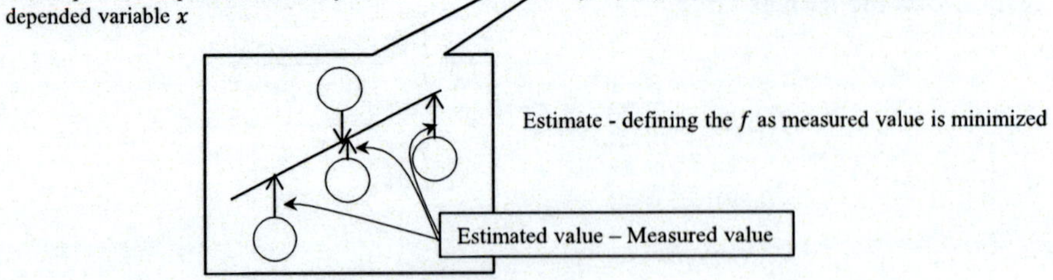

Fig. 4.22 Linear regression in a two-dimensional plane

$y = f(\boldsymbol{x})$ is found statistically using past data. At this time, y is called the independent variable, x is the dependent variable, and the function f is the regression equation.

In the estimation of the FastADR baseline, we use the time series of power consumption, data on weather (outside temperature, humidity, wind speed ...), etc. as dependent variables. By using the regression method as such various past and present dependent variables, it is possible to improve the estimation accuracy of the baseline in short time slots such as minutes.

In baseline estimation, the linear form is often used for regression equations to estimation of the power consumption [4.26, 4.27].

$$y = \boldsymbol{w} \cdot \boldsymbol{x} + \varphi \tag{4.9}$$

Here, $\boldsymbol{w}$ is a weight applied to each dependent variable and is called a regression coefficient. φ is a constant term and an intercept (point where a graph crosses one of the Cartesian coordinate axes) of y.

The regression coefficient and the intercept are set so that the difference between the estimated value and the measured value in the regression equation is minimized, as shown in the lower part of Fig. 4.22. Specifically, the method of least squares is often used. And for nonlinear regression, a baseline estimation method based on Gaussian Process Regression [10].

On the other hand, the pattern matching method is a method of classifying the state before and after the time zone in which the demand response occurred, finding a day with similar pattern from the past data of non-DR day, and using it as a baseline as in Fig. 4.23. Therefore, the method uses more data around the demand response time zone than the regression method. There are various ways to determine this pattern, but many use machine learning [14]. In addition, it is also called data-driven modeling to use data directly [19].

Clustering and Neural Networks are the main machine learning methods used in pattern matching baseline estimation. Clustering is a data classification method that classifies data into groups called clusters. The similarity between the data is calculated, and the data with high similarity is put together in one cluster. Clustering is the type of "unsupervised learning". And unsupervised learning is one of type of machine learning. Unsupervised Learning does not have labeled outputs, so its goal is to infer the natural structure present within a set of data points. The neural network described below is based on supervised learning, where learning and classification are separated. Supervised Learning is done using a ground truth, i.e., we have prior knowledge of what the output values for our samples should be. Hence, the goal of supervised learning is to learn a function that, given a sample of data and desired outputs, best approximates the relationship between input and output observable in the data. Broadly speaking, the above regression equation is also supervised learning.

K-means method [19] is a typical algorithm for clustering. The K-means method is a heuristic algorithm that determines the cluster centers. So that the distance between each data and the cluster center to which the data belongs is minimized when the data are classified into K-clusters, as shown in Fig. 4.24. Data classification (left) and cluster center update (right) are alternately repeated.

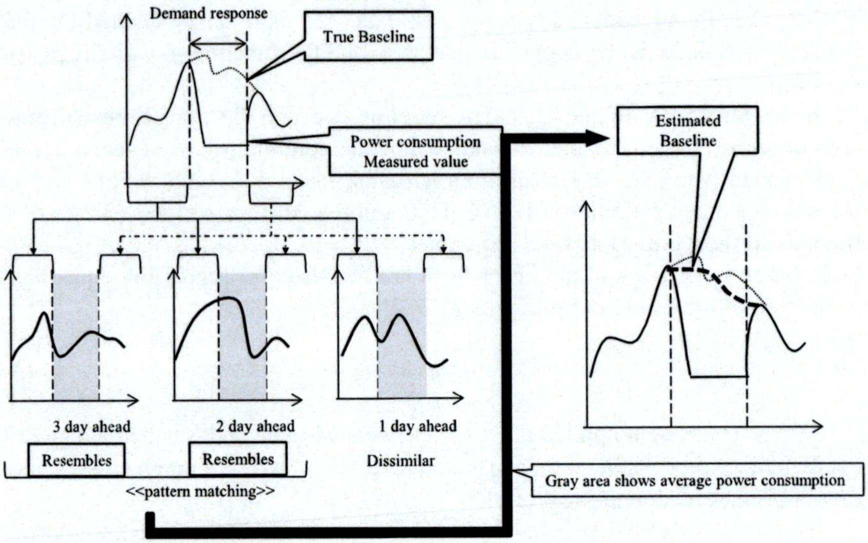

Fig. 4.23 Conceptual diagram of pattern matching method

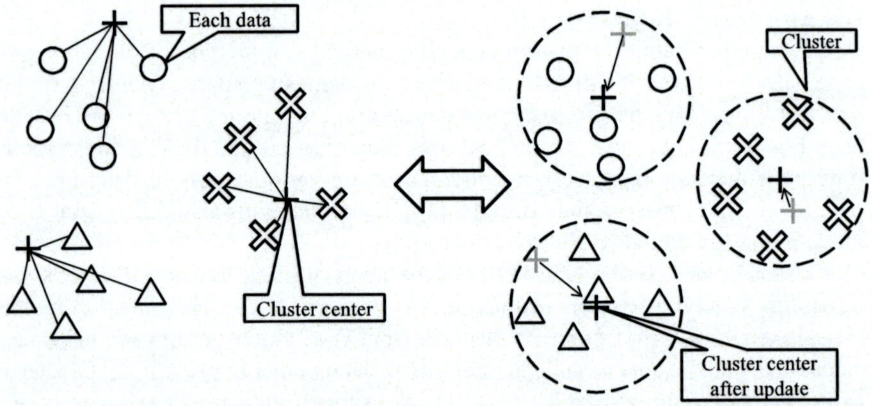

Fig. 4.24 Conceptual diagram of clustering

A neural network is a supervised learning model that simulates the function of human brain cells, and has an input layer for inputting data, an intermediate layer with multiple neurons, and an output layer for outputting results. Figure 4.25 is a conceptual diagram of a neural network, where circles are input variables, rounded-squares are neurons, squares are output layers, and triangles are outputs. These are hereinafter collectively called Nodes. Connect the nodes of the input layer and the intermediate layer, connect the intermediate layer and the output layer, weight the input from the previous layer, sum up, and output to the next layer through a function called Activation Function. At this time, it is called the universality theorem [20] of

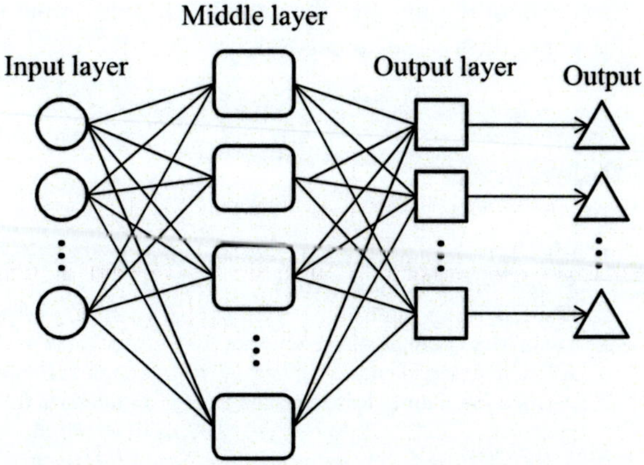

Fig. 4.25 Conceptual diagram of neural network

neural networks that it can be approximated to any continuous function by increasing the number of neurons in the middle layer. Neural networks are highly accurate for regression of complex functions and pattern classification of data with many inputs. Details of model construction by neural network will be described later.

A data mining approach [19] has been proposed as baseline estimation by pattern matching using the machine learning method as described above. This is a combination of Self-Organizing Maps (SOM), which is a type of neural network, and K-means clustering.

As described above, various methods are considered to accurately estimate the baseline of FastADR in demand response. It should be noted that these methods estimate baselines statistically based on past data, so it cannot be estimated sufficiently that there is a small amount of data. That is, in order to improve estimation accuracy with these methods, a large amount of data is required. This is particularly necessary in pattern matching method. It is also important to preprocess the data, sort the necessary data, for process them easily.

According to the explanation in this section, does the condition after the end of the demand response really agree between the case without the demand response and the case without it?

In fact, the results of baseline estimation are changing due to rebound phenomena under the impact of demand response. In rebound period, a device that has been suppressed by demand response releases more power than before. Such that a spring that has been held down extends by inertia after cancellation of the suppression. For example, in air-conditioning, if you try to lower the power consumption by demand response and rapidly lower the room temperature immediately after the end of the demand response, the power consumption may be higher than before the demand response.

This rebound phenomenon is very troublesome, and not only affects the baseline estimation, but also consumes the negawatts generated by the demand response on its own. Therefore, it is also necessary to control the movement of equipment such as air-conditioning after the demand response time so that no rebound occurs after the generation of negawatts.

References

1. IEEJ JEC Technical report, *Standard Specifications for Energy Service by Storage Battery System JEC-TR-59002* (2018)
2. JPEX, Trading guide. http://www.jepx.org/outline/pdf/Guide_2.00.pdf
3. Japan Meteorological Agency. http://www.jma.go.jp/jma/menu/report.html
4. JPEX, Transaction data. http://www.jepx.org/market/index.html
5. Agency for Natural Resource and Energy, Guidelines for negawatt trading. https://www.meti.go.jp/press/2020/06/20200601001/20200601001.html
6. J. Eto, *Demand Response Spinning Reserve Demonstration* (Lawrence Berkeley National Laboratory, 2007)
7. OpenADR Alliance, *OpenADR 2.0 Profile Specifications* (2013). http://www.openadr.org/specification-download
8. IEEE standard, *Ubiquitous Green Community Control Network Protocol* (IEEE1888, 2011). http://standards.ieee.org/standard/1888-2011.html
9. K. Ma, G. Hu, C. Spanos, Distributed energy consumption control via real-time pricing feedback in smart grid. IEEE Trans. Control Syst. Technol. **22**(5), 1907–1914 (2014)
10. J. Morikawa, T. Yamaguchi, C. Ninagawa, Smart grid real-time pricing optimization management on power consumption of building multi-type air-conditioners. IEEJ Trans. Electric. Electron. Eng. 11, pp. 823–825 (2016)
11. J. Morikawa, C. Ninagawa, Fast real-time pricing optimization control of power consumption and room temperature for building multi-type air-conditioning facilities. IEEJ Trans. Power Energy **136**, 817–823 (2016). ((in Japanese))
12. Y. Aoki, H. Ito, C, Ninagawa, J. Morikawa, T. Inaba, S. Kondo, Complex coordination control with entire power limitation and partial temperature priority of building multi-type air-conditioning for real-time pricing. IEEJ J. Indus. Appl. **138**(10), 777–786 (2018) (in Japanese)
13. C. Ninagawa, Y. Aoki, J. Morikawa, T. Inaba, S. Kondo, H. Otake, Area complex adaptive control of building multi-type air-conditioning facilities for real-time pricing. Soc. Heating, Air-conditioning Sanitary Eng. Japan **260**, 23–30 (2018). ((in Japanese))
14. Y.M. Wi, J.H. Kim, S.K. Joo, J.B. Park, J.C. Oh, Customer baseline load (CBL) Calculation using exponential smoothing model with weather adjustment, in *Transmission and Distribution Conference and Exposition: Asia and Pacific, T and D Asia* 2009 (2009)
15. Y. Weng, R. Rajagopal, Probabilistic baseline estimation via Gaussian process, in *2015 IEEE Power & Energy Society General Meeting*, pp. 1–5 (2015)
16. F. Wang, K. Ki, C. Kiu, Z. Mi, M.Shafie-khah, Synchronous pattern matching principle based residential demand response baseline estimation: Mechanism analysis and approach description. IEEE Trans. Smart Grid, 1–13 (2018)
17. T.K. Wijaya, M. Vasiriani, K. Aberer, When bias matters: An economic assessment of demand response baselines for residential customers. IEEE Trans. Smart Grid **5**(4), 1755–1763 (2014)
18. K. Coughlin, M.A. Piette, C. Goldman, S. Liliccote, Statistical analysis of baseline load models for non-residentital buildings. Enery Buil **41**(4), 374.281 (2009)
19. S. Park, S. Ryu, Y. Choi, H. Kim, A framework for baseline load estimation in demand response: Data mining approach, in *2014 IEEE International Conference on Smart Grid Communications*, pp. 638–642 (2014)
20. M. Behl, A. Jain, R. Mangharam, Data-driven modeling, control and tools for cyber-physical energy systems, in *ACM/IEEE 7th International Conference on Cyber-Physical Systems (ICCP)*, pp. 1–10 (2016)

Chapter 5
Battery Control in Virtual Power Plant

5.1 Supply and Demand Adjustment by Storage Battery System

5.1.1 Battery System Control in Virtual Power Plant

According to the literature, the power batteries can be broadly divided into the following three categories as their usage. The first is a system frequency fluctuation suppression measure, which has been used mainly in thermal and hydroelectric power plants, but in recent years, demand and supply adjustment with a fast responsive power storage battery system has come to be expected. Secondly, it is used as a measure against surplus power, and surplus power is generated during the daytime due to the recent large-scale integration of photovoltaic power generation. In order to avoid the wastage of such natural energy, it is useful method that charges the battery with surplus energy like a pump in a pumped storage hydropower plant. Third, it is used as a virtual power plant, which is the theme of this book. Charge and discharge capacity of batteries installed in a distributed manner is centrally controlled by advanced ICT network technology. This is a convenient method that controls the charge and discharge power as a single power plant by remotely controlling the grid-connected power conditioner of the power storage battery system group.

Virtual power plant has several types of different functions. One thing, the power storage battery system itself serves as the power generation source for the virtual power plant, and provides kWh as an integrated power value for a certain duration. In this category, there is a common practice to use the battery system as slow demand response in recent years, but basically, the goal was to store a large amount of power rather than a fast charge/discharge response.

The other is function that compensates for negawatt generated by customer loads such as the building multi-type air-conditioner group FastADR. This function compensates the FastADR response by the instantaneous power kW that charges and

C. Ninagawa, *Virtual Power Plant System Integration Technology*, Power Systems,
https://doi.org/10.1007/978-981-16-6148-8_5

discharges the battery system at high speed in order to make the negawatt generation accurately followed by the supply and demand adjustment.

In 2019, the first function is in the practical stage and the latter function is in the research stage. This section describes storage battery control for selling the large amount of kWh. One such application is described in the previous chapter of this book, where the 30 min accumulated energy is bid for the wholesale electricity market. There is an example of JEPX one-hour-ahead market [1], where one day is divided into 48 frames in 30 min units, and an attempt is made to trade at a market price that balances power supply and demand at least one hour before the execution time.

In addition, the FastADR itself is in the research and development stage for the compensation function for supply and demand adjustment signal tracking of FastADR Negawatt generation by instantaneous charging and discharging. And there is no standard that defines a detailed control method. This book assumes high-speed negawatt virtual power generation by the building multi-type air-conditioner group FastADR that receives a supply and demand adjustment command at every 5 min.

However, if the aggregation of the FastADR of large number of building multi-type air-conditioning units is preformed, the uncertainties of the aggregated virtual power generation will be mitigated by the averaging effect. In this case, after releasing the power suppression command, rebound will occur due to consuming the more power by air-conditioning units for settling the room temperature up to comfortable level.

Due to rebound phenomena, air-conditioners power consumption pattern could not match accurately with supply and demand adjustment signal. Therefore, it is necessary to compensate the difference between the target power of the supply and demand adjustment signal and the FastADR negawatt power. And it is possible using high-speed charge/discharge characteristics of the battery system. Section 6.4 describes a simulation that examines the compensation of the FastADR power response waveform of a building multi-type air-conditioning units by charge/discharge control of such a high-speed battery system.

In control of battery power described in Sect. 5.2 below, the resource aggregator near the customer facility handles the supply and demand adjustment command that is sent to and received from the customer BEMS or ERC. The aggregation coordinator at the top of the entire virtual power plant receives a supply and demand adjustment command that will be transmitted from the power grid to the power plant.

The aggregation coordinator distributes the supply and demand adjustment signal from the grid operation organization to the subordinate resource aggregator. At this stage, it is considered that information is transferred from the aggregation coordinator to the resource coordinator according to the OpenADR communication standard.

The resource aggregator is responsible to propagate the negawatt generation request to the subordinate customer's BEMS by the indirect control method using the OpneADR communication standard. However, the direct control method is adopted for generation the negawatts, when the subordinate customer of the resource aggregator installs the ERC instead of BEMS and the data storage system shall be controlled by transmitting data according to the IEC 61850 communication standard.

In the following sections, the exchange of supply and demand adjustment command for indirect control and direct control for battery systems will be described in detail.

5.1.2 Specific Supply and Demand Adjustment Control of Power Batteries

Now, the compliance ensurance must be provided to individual consumers who's actually prepare a power storage battery system and make it a component of a virtual power plant for supply and demand adjustment. Figure 5.1 shows the system configuration for realizing energy services using a battery system.

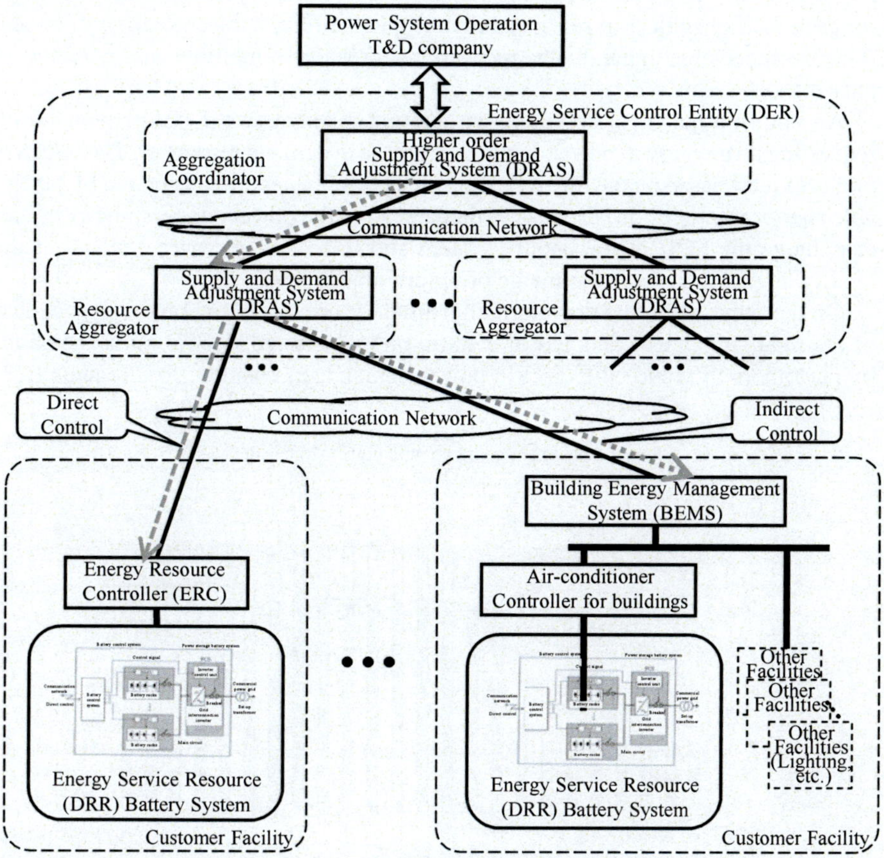

Fig. 5.1 Conceptual diagram of SOC of storage battery (lithium ion battery etc.) (The Institute of Electrical Engineers of Japan: JEC-TR-59002: Standard Specification for Energy Service with Battery System (2018), Partially Modified from p. 13)

Battery system control methods can be broadly divided into two types: indirect control based on supply and demand adjustment commands that indicate only the power adjustment. And direct control over the devices that involved in battery system.

In indirect control, an adjustment signal propagate between the energy service provider's (resource aggregator) supply and demand adjustment system and the customer's building energy management system (BEMS) to control the power demand. The power demand adjustment command from the energy service provider's supply and demand adjustment system is same for both the building energy management system (BEMS) and the energy resource controller (ERC). The reception/transmission system on the customer facility side is different depending on the choice of the customer.

In addition, the result of energy service is based on the measurement of a trading watt-hour meter, which is an interface in the power sales contract between the customer and the electric power company. In addition, when there are multiple facilities subject to energy service in one facility, it is allowed to install a watt hour meter for each facility subject to service, and multiple facilities subject to energy service. It is also acceptable to handle the measurement results of multiple energy meters at once.

An upturn adjustment service is an attempt to increase the power demand by requesting an increase in power demand (increase in demand response). The issuance of this request based on the forecast of surplus power generation generated by renewable energy resources. In response to demand from the power supplier, the customer can charge the SOC of the battery system and increase the power demand of the facility. Figure 5.2 shows the image of upturn adjustment.

On the other hand, the downturn adjustment service is based on the expectation of the tight supply and demand of electric power, and the electric utility demands by customers to reduce the electric power demand. In response to a request from

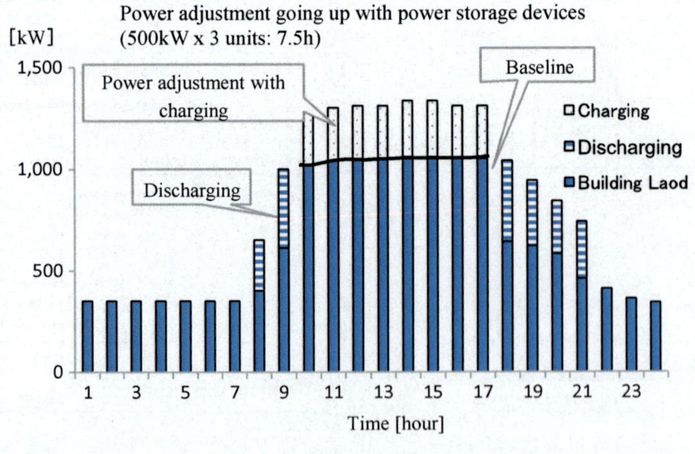

Fig. 5.2 Image of up going adjustment operation by battery system

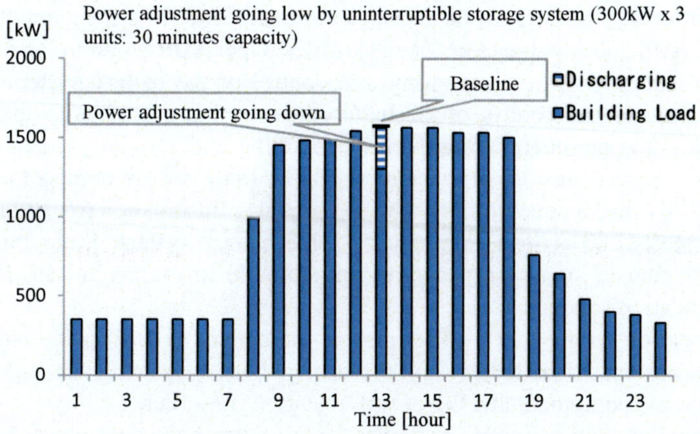

Fig. 5.3 Image of lowering adjustment operation using battery system

an electric power company, the customer discharges the SOC of the storage battery system, if there is a margin, reducing the power demand of the facility. Figure 5.3 shows an image of downturn adjustment. In this way, the electric utility can cope with the shortage of power supply, and the customer get incentives by cooperating with the reduction of power demand.

Electric utilities tries to reduce their power generation by suppressing their power generation against surplus power. However, if the consumer increases the demand and consumes surplus power through the upturn adjustment service, the power supply/demand balance can be maintained and battery system can be charged at a lower price.

Consumer operates the battery system according to real time electricity tariff. In other words, when the electric tariff is low, the battery system can be charged and discharged when it is high.

5.2 Supply and Demand Adjustment Service Using Battery

5.2.1 Service Information of Battery Storage System

The types of information used in supply and demand adjustment service are summarized below. JEC-TR-59002 recommends OpenADR for indirect control and IEC 61850 for direct control as communication protocol standards used to exchange control information.

Information on the amount of power for power demand adjustment: Information indicating the power demand adjustment request of a specific power amount sent from the supply and demand adjustment system to building energy management system or energy resource controller of consumer. Since it is the amount of power to be adjusted and suitable to exchange with the OpenADR communication service.

Information on status monitoring and control of the battery system: Information on monitor and control of the battery system that is exchanged between the battery and the peripheral equipment via building energy management system or energy resource controller of consumer. The building energy management system or the energy resource controller converts the power information for demand adjustment requests into control commands for the battery system. Since this is information related to supervisory control and suitable to exchange with IEC 61850 communication service.

Information on the state and settings of the battery system: Information on the state and settings of the battery and the storage battery system exchanged between the peripheral equipment and the energy resource controller.

Information on the amount of power for confirming the consequences of power demand adjustment: Information obtained by measuring the result of adjusting the power demand of the battery system. Figure 5.4 shows the relationship between these types of information.

When the energy management entity, that is, the supply and demand adjustment system, determines that the power demand needs to be adjusted, it sets up an energy service operation schedule. Then, it notifies the contracting customer and confirms whether or not it is possible to implement energy services such as power demand adjustment. For example, request one-day-ahead/one-hour-ahead/10 min—ahead requesting demand adjustment. This demand adjustment request is called demand adjustment event notification.

And when it is confirmed that the apparent demand adjustment based on the amount of power is done, the incentive determined by the pre-contract is paid to the customer. This is called advance notice type energy service.

Figure 5.5 shows the downturn adjustment service by indirect control. Here, the communication sequence is based on the use of the OpenADR 2.0b profile specification [2]. The energy service control entity contracts with multiple customers and manages the battery system of multiple customers in order to provide energy services. When the energy service is operated, the measurement is notified to the consumer as a demand adjustment consequence, and the consumer is notified from the energy service control entity.

Figure 5.6 shows the downturn adjustment service by direct control. The communication sequence here is based on the use of the OpenADR 2.0b profile specification. In direct control, remote control is performed from the energy service control entity to the customer's storage battery system. Notify the energy service control entity from battery power measurement status after the downturn adjustment operation. And, according to the electric tariff information provided by the electricity supplier, the consumer optimally controls the power demand of the customer's equipment by performing charge/discharge control of the storage battery system.

When supply and demand of electric power becomes tight, the energy service control entity notifies the customer about supply and demand adjustment request

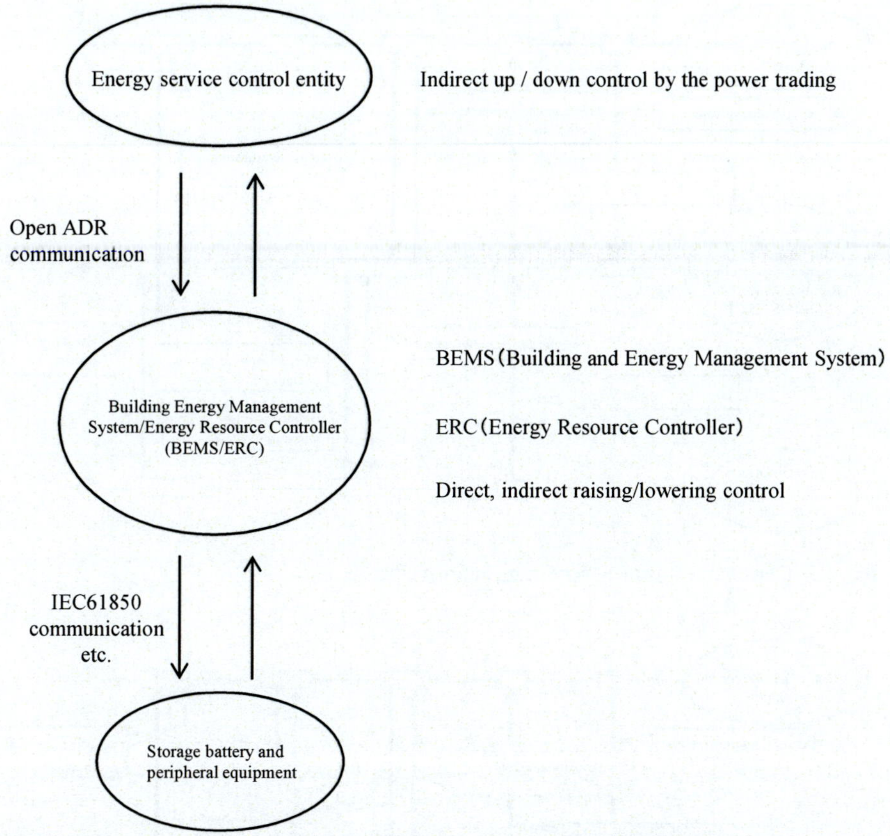

Fig. 5.4 Energy system actor-to-actor exchange with storage battery system

(energy service operation schedule etc.). The consumer then returns demand adjustment compliance (participation, non-participation, etc.) to the energy service control entity. Advance notice of demand adjustment is made at intervals of 10 min/one hour/one day ago. Based on the demand adjustment request, Fig. 5.6 shows the down adjustment service by direct control for the storage battery system. The different of aforementioned indirect control downturn adjustment service is the sequence of "remote control of battery system" and "downturn adjustment operation". Notify the energy service control entity from the storage battery system of the state of battery power measurement after the down adjustment operation. And, according to the electricity price information provided by the electricity supplier, the consumer optimally controls the power demand of the customer's equipment by performing charge/discharge control of the storage battery system.

In Fig. 5.7, regardless of the energy management entity, the electric utility obtains the electricity rate presented by the (retailer) and exchanges information for performing the optimal control of the battery system.

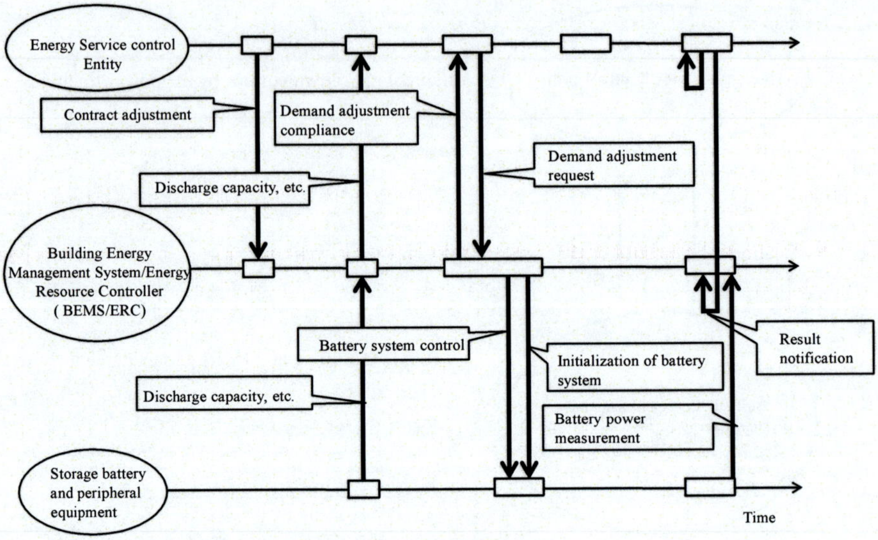

Fig. 5.5 Downturn adjustment service by indirect control

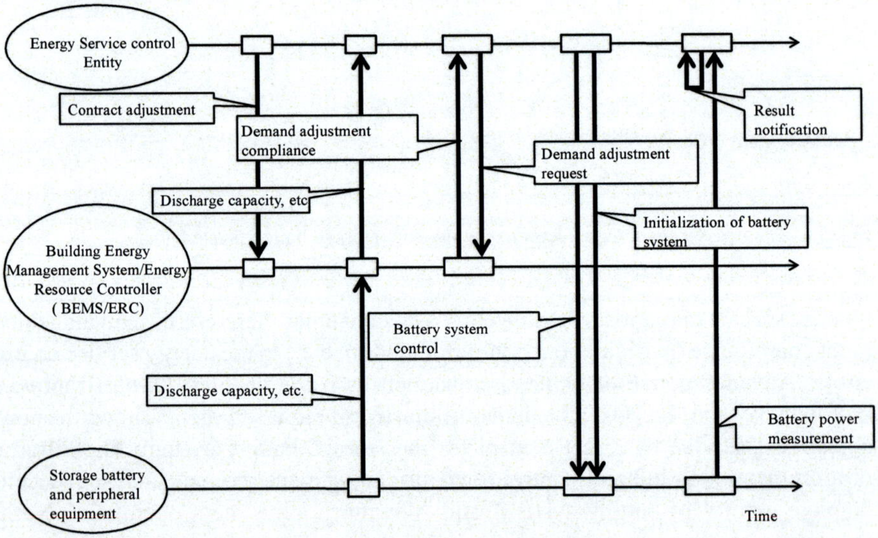

Fig. 5.6 Downturn adjustment service by direct control

Figure 5.8 shows a model of information exchange when the energy service control entity directly controls. According to the electricity tariff information provided by the electric utility, the consumer battery system is directly controlled, so it is possible to optimally control the electricity demand at the customer facility.

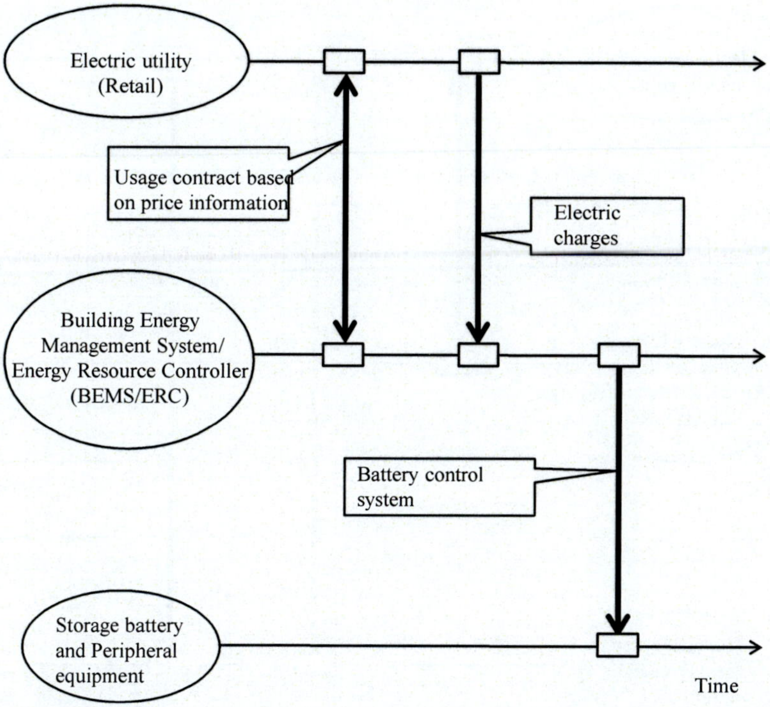

Fig. 5.7 Information exchange for optimal use of electricity based on electricity price information

5.2.2 Electricity Demand Adjustment Service Command

Here, the contents of the data included in the power demand adjustment service command are explained.

Demand adjustment request information based on power consumption is shown in Table 5.1. It consists of demand adjustment request information based on the electric energy from initial time of demand adjustment, continuance (schedule), and adjusted electric energy during that period.

The customer can respond to accept or reject this demand adjustment request. Even after acceptance, demand adjustment can be stopped by paying a penalty for the convenience of the customer.

In addition, supply and demand adjustment results are defined as information, acquisition signal from battery control system, setting signal to battery control system, acquisition signal of storage battery status from BMU (Battery Management Unit) for each storage battery rack, battery storage to BMU transfer signals for energy services, such as setting signals for control, acquisition signals from power conditioners, and setting signals to power conditioners.

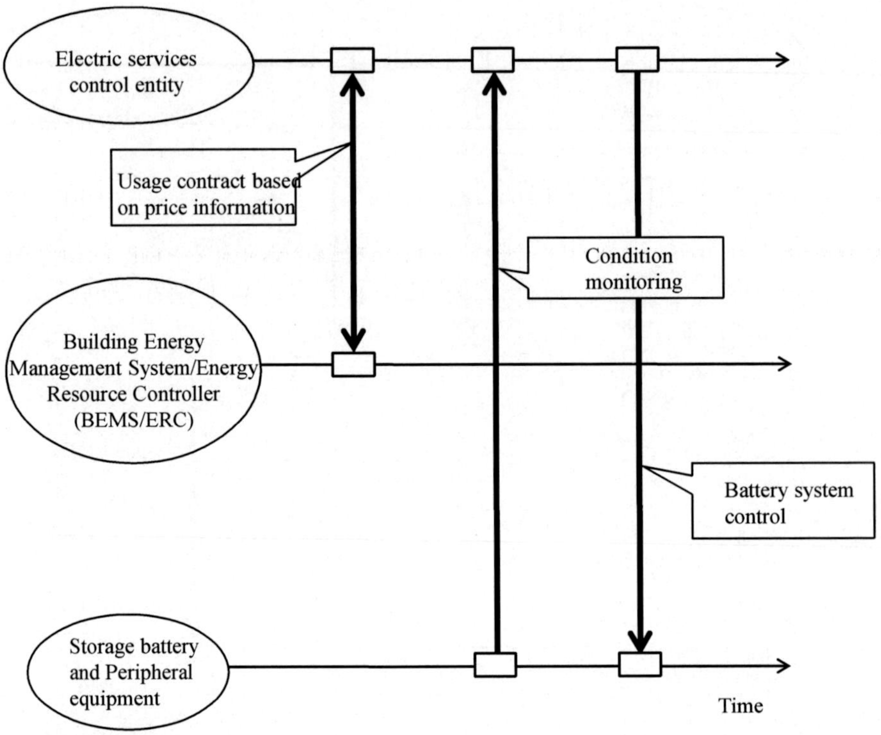

Fig. 5.8 Information exchange for optimal use of electricity based on electricity price information

Table 5.1 Demand adjustment request information

Item name	Description	Unit (type)
Demand adjustment start date	Date and time when demand adjustment starts	(Date time)
Need to adjust the agency	Period of continuing demand adjustment	(Duration)
Period from request notification to start time	The period from the notification date and time of the demand adjustment request to the date and time when adjustment is actually started	(Duration)
Need to adjust the artifact equipment	Equipment name subject to demand adjustment	(String)
Need adjustment method	Power value (absolute value/relative value) specification	(String)
Adjustable power unit	Unit/Scale	W/k
Adjustment power value	Adjustment power value (absolute value/relative value)	(Numerical)
Facilities subject to demand adjustment	Name of facility (customer) subject to demand adjustment	(String)

5.2.3 Communication Service According to IEC 61850 Standard

So far, the IEC 61850 series has been described as a power system standard, but now the IEC 61850 series standard is used in communication networks for distributed power sources such as solar power plants and wind power plants mainly in Europe. IEC 61850 was originally developed as a communication between power devices that require high-speed control in substations.

In mapping services to IEC 61850 MMS (ISO9506) [3] and TCP/IP and Ethernet, the communication service used for control of battery systems according to present version of JEC-TR-59002 is a real-time control over TCP/IP. It is designed to withstand high-speed control communications such as circuit breakers and storage battery modules in power storage battery systems. Here, message class 2 (within 100 ms) is required for communication delay time as categorized in Table 5.2. In addition, the measurement time of the measurement data and the event occurrence time are added to the data as time stamps. SNTP (Simple Network Time Protocol) [4] is used for time synchronization between devices and equipment. And the time synchronization accuracy (time stamp time) between devices and equipment is assumed to be within 1 ms. For this reason, equipment protection functions such as high-speed circuit breakers and instantaneous input of current and voltage require responses faster than 1 ms, and must be performed at lower-level Ethernet level communication.

In addition, security for power system is also possible using IEC 62351 [5]. TLS (Transport Layer Security) in TCP is endpoint authentication, and the IEC 61850. ACSI (Abstract Communication Service Interface) communication service establishes a session between devices by mutual authentication using tokens such as digital certificates and signatures. This is a handshake in which the session key is

Table 5.2 IEC 61850 communication data transfer time specification [Created from IEC 61850-5 (2003)]

Message class	Message type	Time [ms]	Project
1	Fast message	3	Trip, Block, …
		10	Release, Status, changes, …
		20	High automatic compliance
2	Medium speed message	< 100	Low automatic compliance
3	Slow message	< 500	Low speed automatic control
4	Original data	3	High sampling
		10	Low sampling
5	File transfer	1000	File, log
6	Time synchronization	1	An event
		0.1	Zero cross
7	Command	500	Control command

passed by the Diffie Hellman key exchange method, and is also used for subsequent data exchange (symmetric encryption algorithm).

Requirement for communication to energy service by battery system is shown in Table 5.3. If a communication error occurs, it is only necessary to complete the retransmission within each allowable transmission delay time.

When IEC 61850 is used for energy services, the communication model Association, GenDataObjectClass, DataSet, BufferedReportControlBlock (BRCB). Each model is shown in Table 5.4. For these, the communication messages required as DataSet are defined in order to control and setting change of the battery system.

The BRCB model of Report service is used to collect measurement values and monitor equipment in IEC 61850 control.

Figure 5.9 shows the relationship between communication service, DataSet, and Logical Node instances when controlling and setting the battery system. For example, in the logical node instances DSTO2 and DINV1, use SetDataSetValues for the corresponding DataSet to start/stop of battery system and for set the active power/reactive power. Control commands are issued on the logical node instance XCBR1 for the on/off of the circuit breakers associated with them (Select, Control).

State changes and measured values are sent to DSTO2 and MMXU1 data objects (DataObject) as control results for the storage battery system and reflected. If the

Table 5.3 IEC 61850 communication requirements for power storage battery control

Application site	Communication service	Cycle	Delay
Communication between energy service control entity and customer system	Supply/demand adjustment request	Anytime (10 min /1 h/ 4 h/1 day before)	1 point 10 points
	Supply–demand adjustment report	As needed (after adjustment)	1 point 5 points
Customer system communication (ERC and xEMS, direct control)	Control/setting	As needed	100 ms About 10 points
	Condition monitoring	As needed	100 ms About 10 points
	Measurement	1 s	100 ms About 20 points
	measurementBMU/PCS	1 s	100 ms 30 points
Storage system communication (ERC/PCS/BMU, electrical signal)	Control/setting	As needed	1 ms About 10 points
	An event	As needed	1 ms About 10 points
	Measurement	1 s	1 ms About 20 points
Time synchronization	Time synchronization accuracy	–	1 ms or less

Table 5.4 IEC 61850 energy service communication for power storage battery control

Model	Service	Function
Association model	Associate	Establish connection from client to server
	Abort	Suspend the client and server connection
	Release	Terminate the connection between client and server
GenDataObjectClass model	GetDataValues	Read data value of DataObject of logical node
	SetDataValues	Write values to all referenced DataObjects
DataSet model	GetDataSetValuesDataSet	Read values of all referenced DataObjects
	SetDataSetValuesDataSet	Write values to all referenced DataObjects
	CreateDataSetDataSe	Creation
	DeleteDataSetDataSet	Delete
	GetDataSetDirectoryDataSet	All included Reading ObjectReferences
BRCB model (Send report from server to client)	GetBRCBValues	Reading BRCB communication settings
	SetBRCBValues	Write communication setting of BRCB
Control model	Select	Selection of control target equipment
	Operate	Control execution
	Cancel	Cancel/Cancel control
	CommandTermination	Control end confirmation notification from server to client

BRCB TrgOps is set to data-change and Integrity, these state changes and measurement values are notified from the storage battery system to the host system through the communication service report.

In the communication service of IEC 61850, communication messages are defined by DataSet. IEC 61850 communication messages can be freely determined by the user as needed using the DataSet function.

In indirect control, information is exchanged between the power system operation system, supply and demand adjustment system, energy resource controller (ERC), or energy management system (BEMS). In the energy resource control device or energy management system, the target device in the storage battery system is monitored and controlled using IEC 61850.

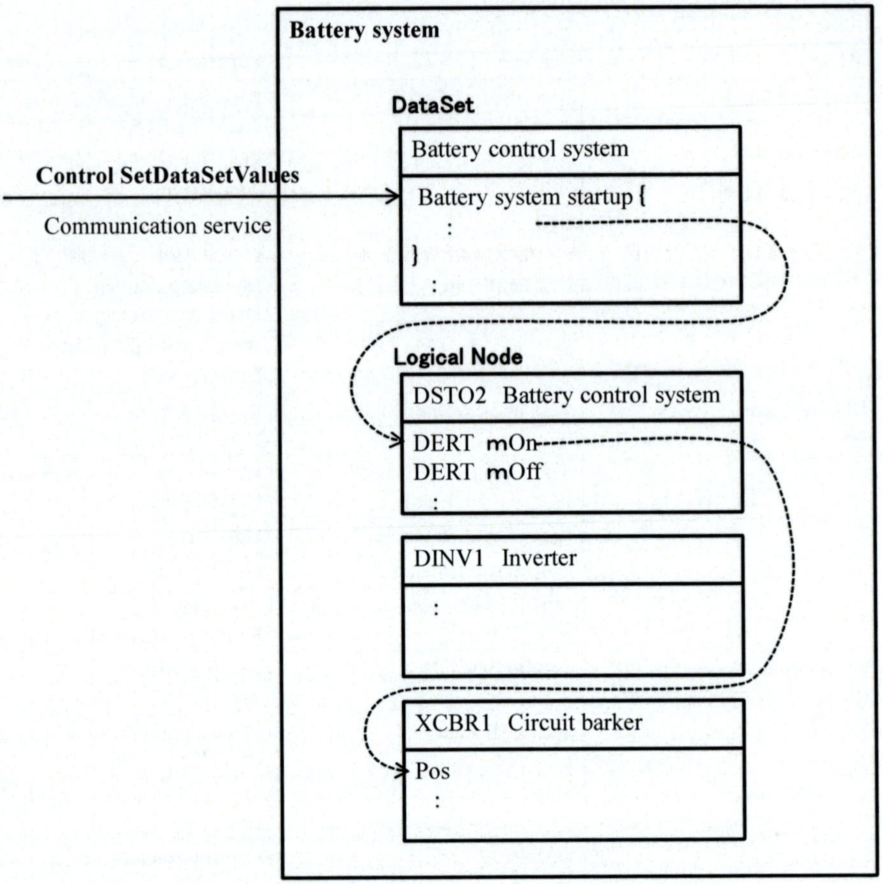

Fig. 5.9 DataSet structure for control and setting to storage battery system

Requests for supply and demand adjustment are made between the power system operation system, the supply and demand adjustment system, and the energy resource control device or energy management system. After that, control according to IEC 61850 is performed in the storage battery system by the energy resource control device or energy management system.

Figure 5.10 shows the range in which IEC 61850 is applied to direct control of energy services using storage batteries. Since the main object of IEC 61850 is monitoring and control of substation equipment and distributed power sources, direct control is suitable for energy services using storage batteries. In addition, the information exchange between the power system operation system of the electric utility and the supply and demand adjustment system of the energy service provider is performed by direct control. Based on the obtained information, the energy service provider's supply and demand adjustment system can remotely control the storage battery system equipment using IEC 61850.

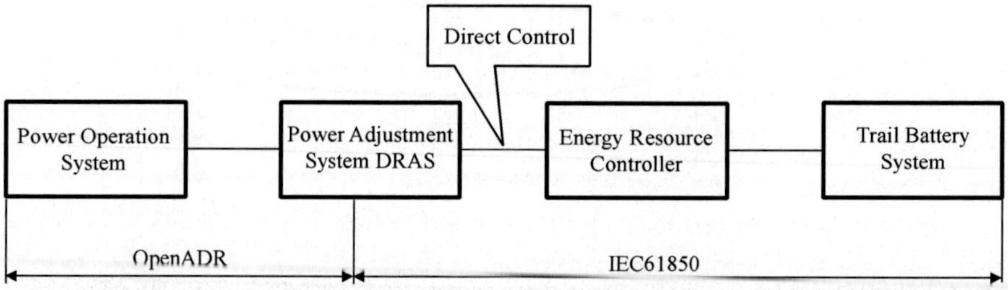

Fig. 5.10 Scope of application of IEC 61850 in direct control service

A request for supply and demand adjustment is made between the power system operation system and the supply and demand adjustment system, and the supply and demand adjustment system controls the storage battery system according to IEC 61850.

In the case of direct control, the supply and demand adjustment system uses the BRCB model report service to monitor the status of the storage battery system as needed. The BRCB stores each DataSet and manages the communication of reports. Use SetBRCBValues to specify the settings for the report service and the start/end of the report. For supply and demand adjustment requests and supply and demand adjustment changes, the supply and demand adjustment system decides compliance based on the operation status of the storage battery system through state monitoring.

The difference in communication sequence between indirect control and direct control is whether the IEC 61850 client is an energy management system, energy resource control device, or supply and demand adjustment system. A request for supply and demand adjustment is made between the power system operation system and the supply and demand adjustment system, and the supply and demand adjustment system controls the storage battery system according to IEC 61850.

In the case of direct control, the supply and demand adjustment system uses the BRCB model report service to monitor the status of the storage battery system as needed. The BRCB stores each DataSet and manages the communication of reports. Use SetBRCBValues to specify the settings for the report service and the start/end of the report. For supply and demand adjustment requests and supply and demand adjustment changes, the supply and demand adjustment system decides compliance based on the operation status of the storage battery system through state monitoring.

The difference in communication sequence between indirect control and direct control is whether the IEC 61850 client is an energy management system, energy resource control device, or supply and demand adjustment system.

5.2.4 IoT Control of Storage Battery System

In development of the virtual power plant, wireless communication using mobile phone lines (Long Term Evolutions: LTE) is attracting attention in order to control

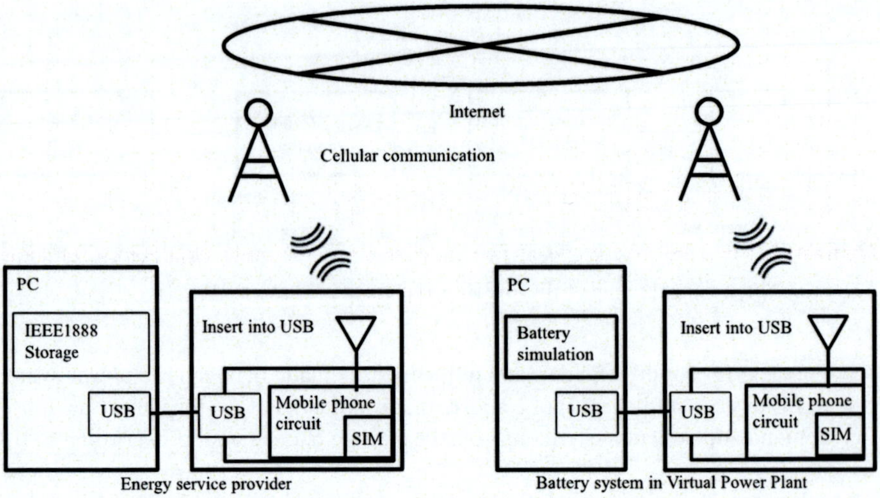

Fig. 5.11 Mobile phone line compatible IoT interface

the battery system. The following is an example of communication features when an energy service provider communicates with a battery system of virtual power plant in an Internet environment.

The experimental facility consists of an energy service provider, two PCs instead of a battery system, and a small communication device (USB dongle) that is inserted into an LTE-compatible USB. LTE is one of the mobile phone communication standards, called 3.9G (4G), and is the mainstream mobile phone high-speed line. Figure 5.11 shows the mobile phone standard's compatible IoT interface.

LTE SIM cards can be obtained by contracting with a major communications carrier. In this experiment, assuming IoT, the monthly usage fee of SIM card is low (instead, the communication speed is limited), and the condition that a global fixed IP address is given, Epsim (WITH Networks Co., Ltd.) was selected which has unlimited communication capacity, communication speed of up to 200 kbps for transmission and receiving, and a global fixed IP address. NTT Docomo's LTE (a Japanese mobile communication provider) line was used in this experiment.

Figure 5.12 shows the transmission of "Ping" transmission data, which is SNMP check communication of the IP network, from the energy service provider's PC to the storage system PC. In testing, a ping was sent every second, and the time taken from sending to response was measured. In most cases, the response time was 200 [ms], but it was widely distributed up to about 500 [ms], and in some cases it took more than 1.0 s. In some cases, there was no response.

Figure 5.13 shows the result of executing the WRITE method on the server using the IEEE1888 protocol, assuming the case where the SOC of the battery system is periodically transmitted to the energy service provider. In this example, the WRITE method actively sends data from the client side to the server side. During experiment, the WRITE method was executed every 10 s, and the time taken from transmission

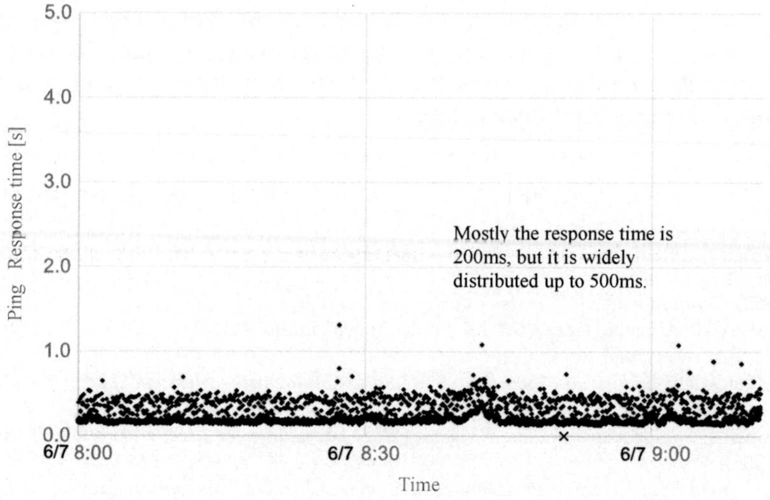

Fig. 5.12 LTE Ping (connection check) response time

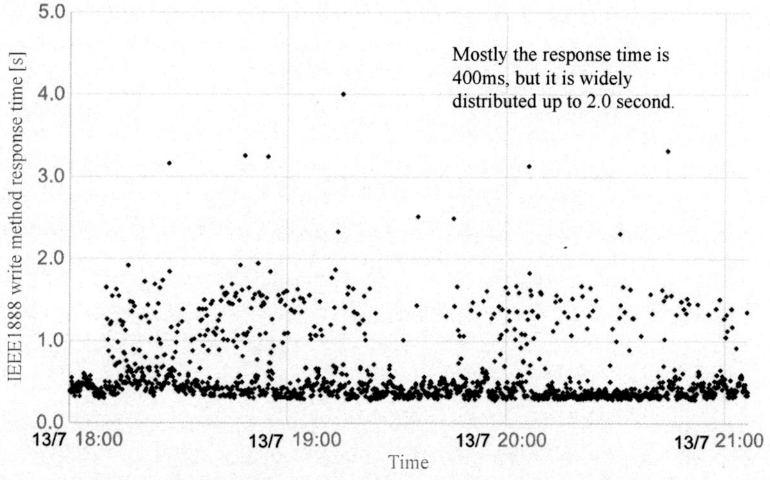

Fig. 5.13 Control command communication response time on mobile phone line

to response was measured. In most cases, the response time is 400 [ms], but it is widely distributed up to about 2.0 s.

As detailed in this chapter, IEEE1888 is an XML-based RPC protocol called SOAP, which uses HTTP as the application layer protocol to transmit messages. For this reason, it has the availability that communication is established in an environment where Web communication is possible. In addition, when there is a communication error such as packet loss, a retry is performed in HTTP. HTTP messages are composed of headers, blank lines, and data. In source code, SOAP XML messages are described in the data part. From the results of these verifications, it seems that using the LTE line of mobile IoT is a good way to control the battery system. This is because even if the line speed is as low as 200 kbps. However, there are many problems of communication packet delay and loss.

References

1. JPEX, *Trading guide*. http://www.jepx.org/outline/pdf/Guide_2.00.pdf
2. OpenADR Alliance, *OpenADR 2.0 Profile Specifications* (2013). http://www.openadr.org/specification-download
3. ISO standard, *Industrial Automation Systems—Manufacturing Message Specification—Part 1: Service Definition*, ISO9506-1 (2000)
4. D. Mills, *Simple Network Time Protocol (SNTP) Version 4 for IPv4, IPv6 and OSI*, RFC4330 (2006)
5. IEC Standard, *Power Systems Management and Associated Information Exchange—Data and Communications Security*, IEC 62351 (2018)

Chapter 6
Demand Side Equipment Control in Virtual Power Plant

6.1 FastADR Response Prediction for Building Multi-type Air-Conditioning

6.1.1 FastADR Response Prediction for Building Multi-type Air-Conditioning

Building multi-type air-conditioning is considered to be promising as a negawatt generation load of FastADR by power control among a number of power customer facilities. There are two reasons for this.

First is the high penetration of building multi-type air-conditioning and the high power consumption. In general office buildings, in case of Japan, the power consumption of air-conditioning accounts for about 50% of the entire building, and in particular, building multi-type air-conditioning is installed in large quantities in small and medium-sized office buildings.

Second, the equipment does not have a serious adverse effect on occupant compared to other consumer equipment (such as lighting).

Now, the following methods can be considered to realize FastADR of building multi-type air-conditioning.

Outdoor unit operation stop: Turn off the outdoor unit and stop the operation.

Pros: It is easy to handle as a control target because there is no uncertain behavior.

Disadvantages: Loss of complete air-conditioning capability can affect occupant comfort.

Change the set room temperature: Change the set temperature of the indoor unit to minimize the power consumption. Pros: Because it is a general function, it can be controlled by any manufacturer. Cons: The set room temperature and power consumption are not a simple relationship and are difficult to handle regarding control targets.

C. Ninagawa, *Virtual Power Plant System Integration Technology*, Power Systems,
https://doi.org/10.1007/978-981-16-6148-8_6

Power suppression command: Directly limits the upper power limit of the air-conditioner. Pros: It is possible to control by directly specifying the amount of power suppression. Cons: Air-conditioning needs to have this function.

Of these, the last method can directly control the power suppression command value and can take into account the effect on building residents. Therefore, the following explains how to implement FastADR. Moreover, it is suitable for the realization of FastADR if suppressing the power consumption from maximum to minimum takes around 10 min [1].

Figure 6.1 shows a conceptual diagram of FastADR aggregation for large number of multi-type air-conditioning units in office buildings. Here, only one aggregator is shown, but in reality, many aggregators contribute to manage a large number of buildings. Each aggregator receives a FastADR total power suppression request from the electric power company. Aggregator manages many building air-conditioners in various buildings and it distributes individual power suppression commands to the air-conditioners in each building. At the time of this distribution, the power suppression command must be properly distributed. So, that comfort in each building is not impaired while meeting the total power suppression requirement of FastADR.

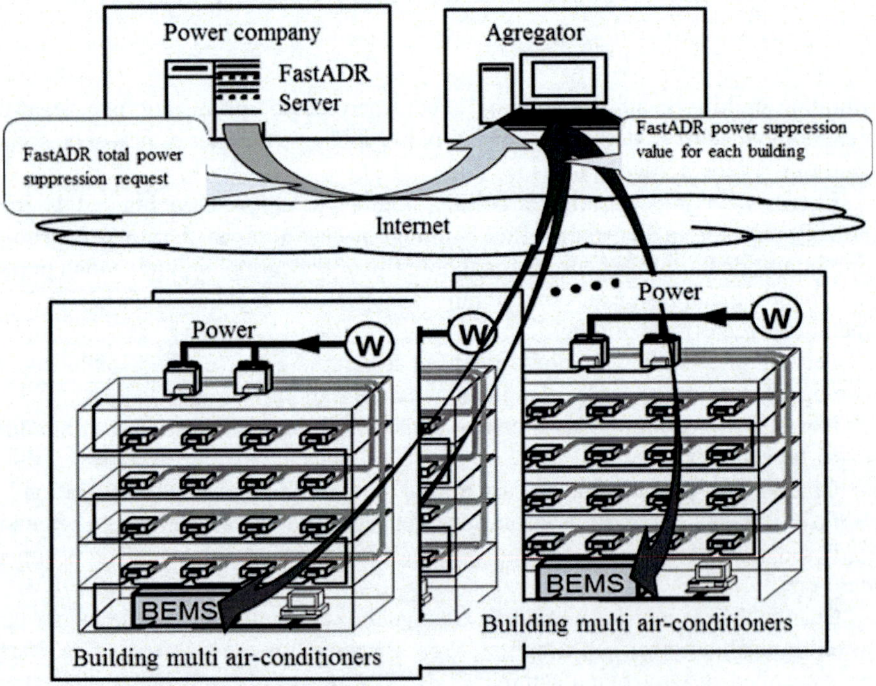

Fig. 6.1 Conceptual diagram of FastADR for building multi air-conditioning

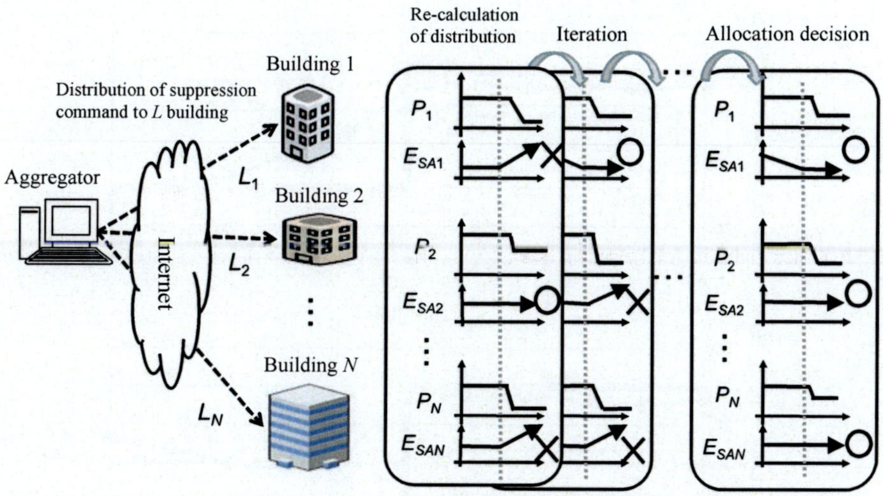

Fig. 6.2 Conceptual diagram of FastADR for building multi-type air-conditioning

Therefore, when distributing the total power control request to each building, the algorithm shown in Fig. 6.2 is assumed. That is, assuming the power suppression command L is distributed to the air-conditioners in each building. And air-conditioning power consumption P and the indoor comfort E_{SA} are predicted as response characteristics. If some air-conditioners are predicted to exceed the allowable value of the room temperature, it is necessary to re-correct and redistribute the power suppression command value to that air-conditioners. And conversely, the air-conditioner that is predicted to still have margin in allowable value of room temperature should be more strongly suppressed. It is considered necessary to have a system control technology that suppresses and redistributes more strongly.

Figure 6.3 shows three examples of the trend of air-conditioning power consumption relative to the power suppression command. In these three examples, the same step-shaped power suppression is commanded, but the response characteristics of the power suppression command received at 7:10 shows that the power consumption is in three different ways. This phenomenon is due to the fact that the power suppression command is different from the target value in the so-called control theory. In other words, the air-conditioning power consumption needs to take into account various factors due to the complex built-in control. So, the power consumption behavior is not uniquely determined by the power suppression command. Therefore, it is necessary to note that the air-conditioning power consumption as the response characteristic of the power suppression command have various stochastic trend depending on the situation. Similarly, the indoor comfort also takes a stochastic trend. This probabilistic phenomenon is thought to be strongly influenced by variations in time series data such as previous power consumption, indoor conditions (room temperature, set temperature, etc.) and weather (outside temperature, solar radiation, etc.). Neural

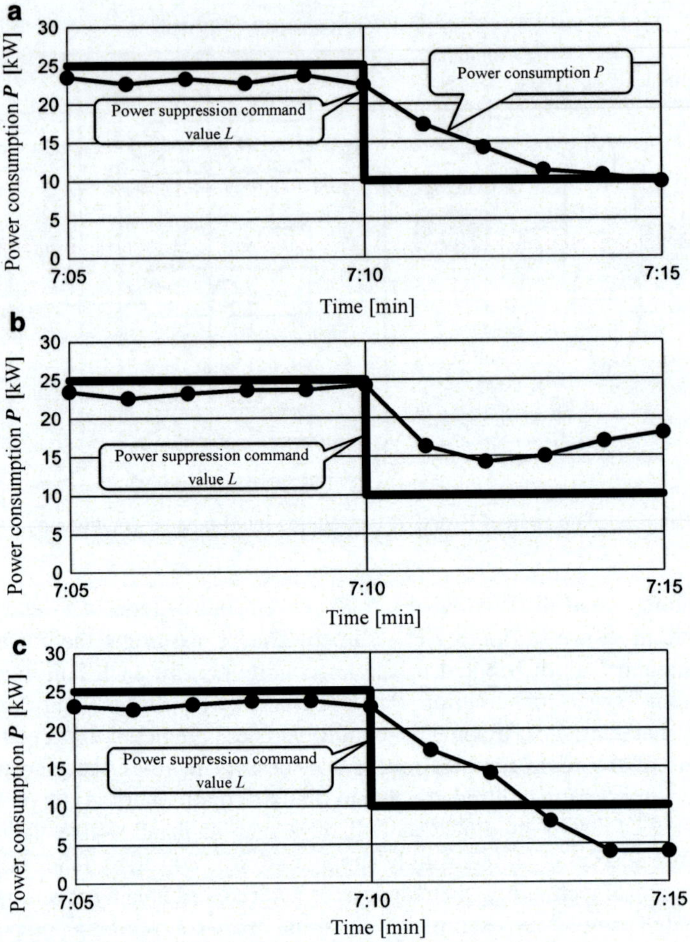

Fig. 6.3 Example of stochastic behavior of air-conditioning power consumption with respect to power suppression command

networks (Fig. 4.25) are suitable for predicting the probabilistic future value from such previous time series data [2–6].

6.1.2 Neural Network Prediction

Neural Networks (NN) are distinguished by being called deep learning because deep learning algorithms are required for four or more layers with two or more intermediate layers. So for, deep learning has been attracting attention as it continuously updates

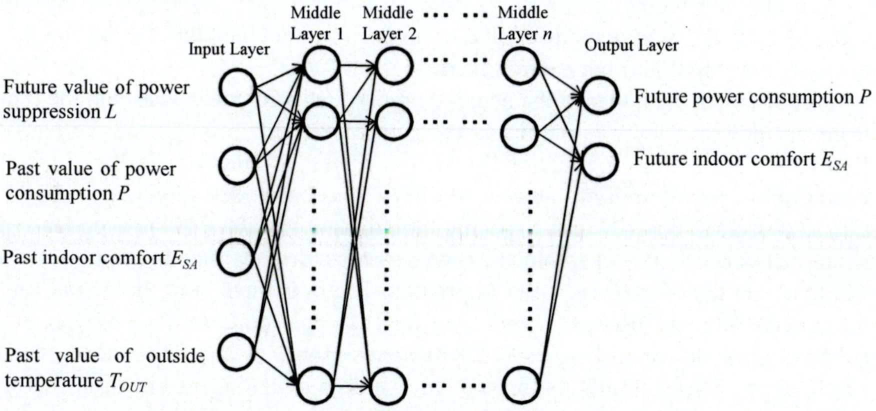

Fig. 6.4 Conceptual diagram of deep learning model

past records in speech recognition and image benchmark tests. Deep learning is also used in object detection and automated driving technologies.

Therefore, this book also describes the creation of a complex and intelligent model of instantaneous power change characteristics of an air-conditioner by deep learning.

Figure 6.4 shows a conceptual diagram of the deep learning neural network structure. That is, the model structure has one input layer, n intermediate layers, and one output layer.

Here, time-series data information is propagated from the input layer to the output layer through the intermediate layer, and output from the left to the right node.

The input/output variables shown in the Fig. 6.4 are only examples. Variables such as the power suppression command value L assumed to be commanded in the future, the previous air-conditioning power consumption P, the past setting and room temperature deviation E_{SA}, the past The outside air temperature T_{OUT} were used as input variables.

On the other hand, as the response characteristics of the result of the power suppression command value L, the future power consumption predicted value P and the future room temperature deviation E_{SA} were used for the output.

Here, the reason why the previous air-conditioning power consumption P and room temperature deviation E_{SA} are included in the input is because the trend before prediction can be learned from the input value. The reason for including the previous outside temperature T_{OUT} in the input is to reflect the state of the external environment in the prediction. In addition, there are useful input variables such as solar radiation and indoor heat load to indicate the state of the external environment, so select them according to the situation. Here, E_{SA} is an index indicating the room temperature deviation, for example, as shown in Eq. 6.1.

$$E_{SA} = T_A - T_{SET} \tag{6.1}$$

T_A is the room temperature and T_{SET} is the set temperature. The larger the absolute value of E_{SA}, the more uncomfortable room environment, and the smaller the E_{SA}, the more comfortable room environment.

In addition, at the nodes of the neural network, including deep learning, the input from the previous layer is output as the sum of weights through the **Activation Function**. This activation function is usually a nonlinear function. Earlier, **Sigmoid Functions** were often used, but recently *ReLU* has been used. Figure 6.5 shows a time series graph describing the specific input/output variables of the deep learning neural networks. In this example, the time axis *t* is a discrete time in minutes. As an example, the data for every 1 min in previous 5 min is input, and the output up to 5 min in the future is shown as output. This is an example, and 5 min is not necessarily the best model for predicting FastADR power response of building multi-type air-conditioners. For example, the reason for 5 min is a significant intention that the microcomputer built-in control of the air-conditioning refrigerant circuit responds to the power control command. This is because the response result with the step change of the power suppression command is not convergent and stable. And the power consumption of the air-conditioner in seconds is irrelevant to the effect of the step response of the power suppression command after several tens of minutes.

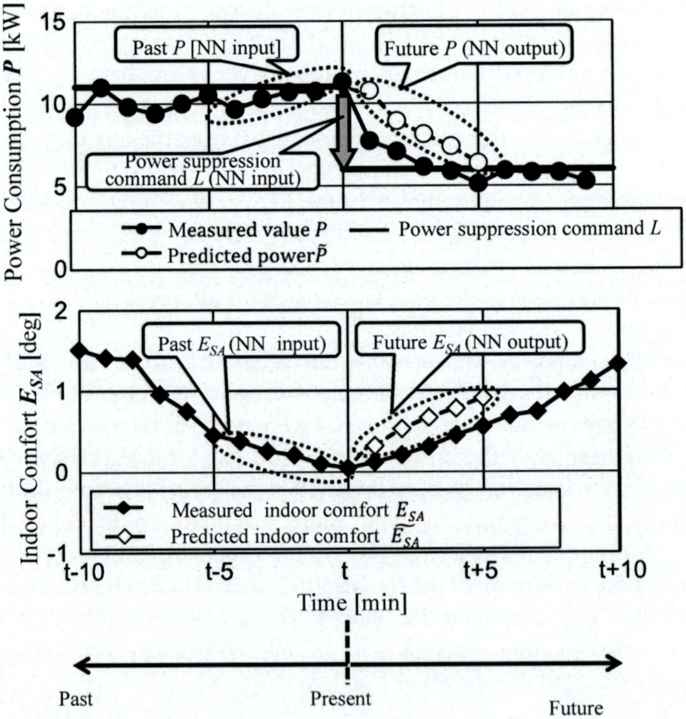

Fig. 6.5 Explanation of deep learning input/output variables

6.1.3 Deep Learning Methods

One of the learning methods in neural networks is the error back-propagation method. This error back-propagation method is what kind of function is approximated by adjusting the weight, etc. However, as in the case of deep learning, increasing the number of intermediate layers may cause vanishing gradient problems. The problem of vanishing gradient is that weights cannot be updated smoothly and learning becomes difficult.

One method of avoiding the vanishing gradient problem is the Stacked Denoising Autoencoder. In this method, the initial value of the weight before updating the weight by back-propagation is not a random number, but the initial value of the weight is determined in advance by prior learning.

Autoencoder (auto-encoder, self-encoder) performs pre-learning that does not have a target output. And restores the input data to the output of the neural network using a training data set. As a result, the initial value of the weight that well represents the characteristics of the individual learning data can be obtained. Figure 6.6 shows the learning model diagram of the autoencoder.

Next, "stacked" means "stacked as it is" and means that autoencoders are stacked from the input layer. Figure 6.7 shows a conceptual diagram of the Stacked Autoencoder. As a 4-layer deep learning neural network, the initial value of the weight W_1 that connects the input layer and the intermediate layer 1 is determined using an auto encoder for the neural network of the input layer and the intermediate layer 1.

Then, the initial value of the weight W_2 connecting the intermediate layer 1 and the intermediate layer 2 is determined using an autoencoder for the neural network of the intermediate layer 1 and the intermediate layer 2. The weight $W3$ that connects

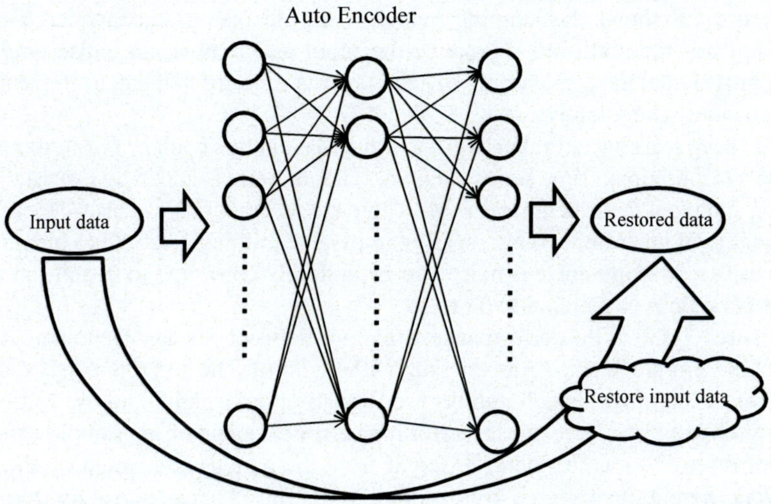

Fig. 6.6 Auto encoder conceptual diagram

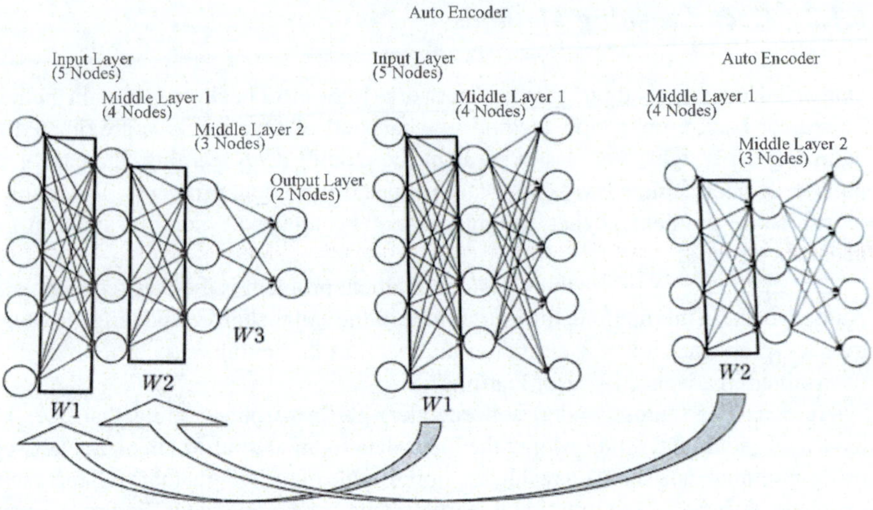

Fig. 6.7 Stacked auto encoder conceptual diagram

the next intermediate layer 2 and the output layer does not constitute an autoencoder, and the initial value of the weight is generated randomly. This mechanism is the same even if the middle layer is dozens of deep learnings. The procedure is divided into layers, the autoencoder is constructed, and the initial value of the weight is repeated into an order.

Finally, Denoising, which means noise removal, is prior learning in which noise is intentionally added. The input data of stacked autoencoder is input, and the original input data is input as output.

Figure 6.8 shows the learning model of the denoising autoencoder. This pre-learning has the ability to reproduce the input and remove the noise, and it has been proven that the pre-learning of each layer is superior to using a simple stacked autoencoder without noise.

The deep learning neural network learning data in this book is a continuous value of the air-conditioner time series data. So, random noise is added by Gaussian distribution and input to the neural network. There are several types of noise, depending on the quality of the training data. As an example, the training data set is a binary signal such as 0/1. Missing noise is noise that is randomly converted to 0, and salt-pepper noise is random conversion to 0 or 1.

Figure 6.9 shows the performance difference between Stacked Denoising Autoencoder and Simple Stacked Autoencoder without noise. The horizontal axis shows the number of learnings (epoch number) for updating the weight, and the vertical axis shows the test error. Here, the test error is an error calculated from sample data (hereinafter referred to as test data) different from the data to be learned by the neural network (hereinafter referred to as training data). The error calculated for the training data is called training error.

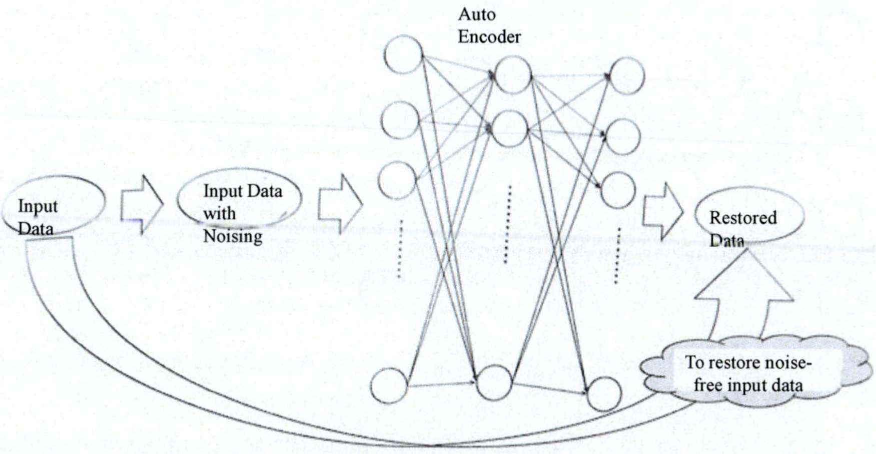

Fig. 6.8 Denoising autoencoder conceptual diagram

Fig. 6.9 Comparison of test error with and without DeNoising

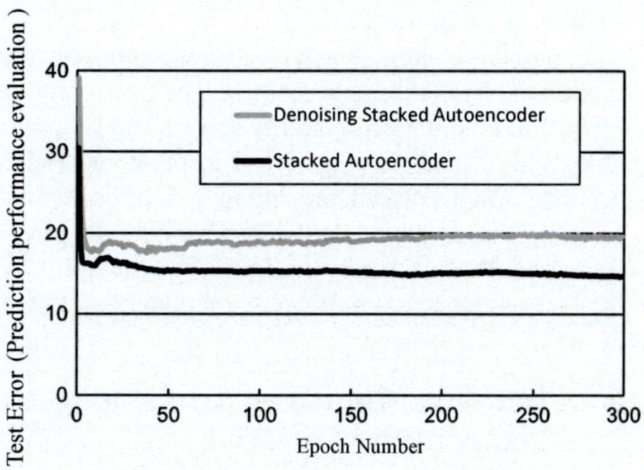

The test error indicates how much prediction performance the neural network has for the data not yet seen. The Smaller value, the better the performance. Figure 6.9 shows that when Stacked Denoising Autoencoder is used, test error is smaller and prediction performance is higher than Stacked Autoencoder.

Also, the dropout method is one of the leading learning techniques in deep learning. This dropout method is a method of learning while randomly erasing neurons in the input layer and the intermediate layer, and has recently attracted attention as a method of suppressing overlearning. Over-learning is a phenomenon in which an error increases when unlearned data is given to a neural network in the same way as training data, even though the error of learning data learned by the neural network is small. In particular, a model with many intermediate layers, such as deep learning, is said to be too fit for the training data set and easy to over-learn.

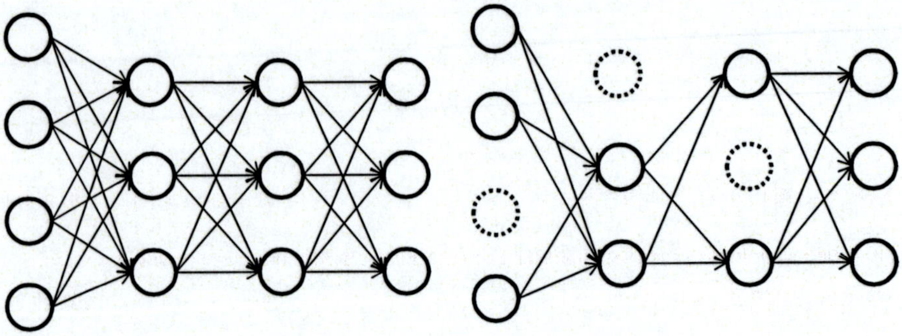

Fig. 6.10 **Left**: Typical neural network with full connection **Right**: Dropped out neural network

In the dropout method, when learning a training data set, the input layer and the middle layer are dropped at a fixed rate as if they did not exist from the beginning. Figure 6.10 shows a conceptual diagram. As a result, the degree of freedom of the neural network is forcibly reduced. The expressive power is lowered, and excessive learning is suppressed.

Figure 6.11 shows examples of training and test errors when over-learning is suppressed by the dropout method. The horizontal axis is the epoch number, and the vertical axis is the error. In Fig. 6.11a, the training error decreases as the number of epochs increases, but the test error increases, indicating that overlearning has occurred. On the other hand, in Fig. 6.11b shows that the over-learning is suppressed by using the dropout method because the test error decreases as the training error decreases.

6.2 Building Multi-type Air-Conditioning FastADR Room Temperature Variation

6.2.1 Neural Network for Predicting the Variation of Room Temperature

Up to the previous section, we described a method for developing a neural network to predict the response of building multi-type air-conditioners to FastADR and the room temperature response. In this section, an example of developing a neural network for predicting the variation in room temperature for FastADR power suppression of building multi-type air-conditioner is described. Neural network is developed by learning from time-series data per minute of a building multi-type air-conditioners installed in an office building. As an example of how FastADR is predicted, the amount of room temperature change after 5 min is predicted, compared with measured data, and prediction accuracy is evaluated. In addition, this section

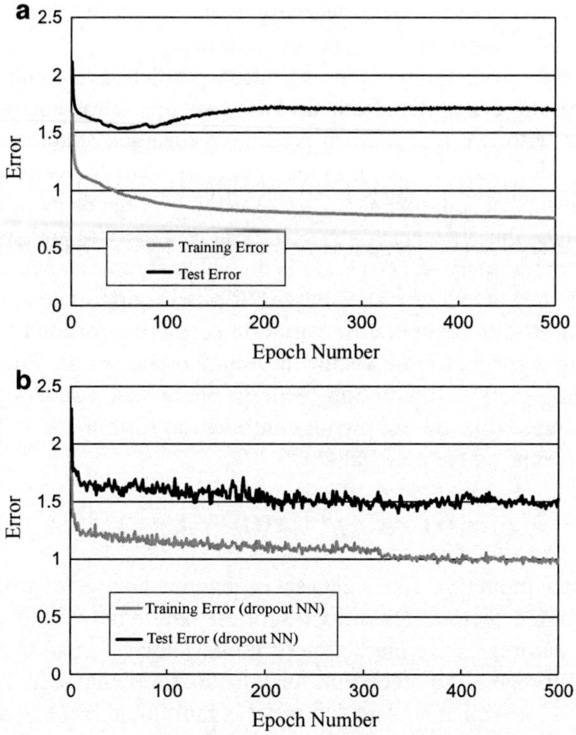

Fig. 6.11 Comparison of training error and test error with and without dropout **a** Without dropout **b** With dropout

describes a FastADR room temperature variation prediction formula based on Linear Autoregression.

Room temperature originally changes slowly, and if it changes for a short time of 5 min, it does not change drastically and nonlinearly. In this example, the average room temperature deviation of the entire target building is considered. So, the range is likely to be supported by the linear autoregressive prediction formula. However, it is necessary to have characteristics that can cope with sudden changes, such as when a large power suppression command is issued during a sudden room temperature change.

Here, as an example, we deal with a practical method that averages and predicts both the linear characteristics and the prediction ability to abrupt changes of the neural network. A simple neural network structure is practically sufficient even if it is not the deep learning of the multi-layered complex structure described in the previous section.

It should not be misunderstood that the power consumption of the building multi-type air-conditioner changes abruptly and complicatedly due to the complex built-in control. Linear autoregressive prediction and a simple neural network predict such type of variations in power. Deep learning as described in the previous section is necessary to predict such type of variation in power.

The FastADR room temperature variation prediction formula treated in this section is an example that includes neural network prediction and linear autoregressive prediction with two independent prediction formulas. Parameters are includes such as; t is the discrete time in 1 min intervals, $L(t)$ is the power suppression value, $P(t)$ is the power consumption at t, $E_{SA}(t)$ is the average room temperature deviation that averages the difference between room temperature and setting at t, $T_O(t)$ is the outside temperature. $E_{SA}(t+t_F)$ is the value after t_F minutes. In this section, $t_F = 5$, that is, the value 5 min after FastADR power suppression start is predicted.

This FastADR room temperature variation prediction formula converts the time series data into a state variable vector in preprocessing stage. The neural network and linear autoregressive prediction formula predict the room temperature index separately and give it as the following combination formula as $E^*_{SA}(t+t_F)$. Here, the symbol * means a predicted value.

$$E^*_{SA}(t+t_F) = \alpha \cdot f^{NN}(\boldsymbol{x}(t)) + (1-\alpha) f^{AR}\left(\boldsymbol{x}'(t)\right) \tag{6.2}$$

In Eq. 6.2, the input time series data set is generated in the preprocessing stage for input state variable vectors $x(t)$ and $\boldsymbol{x}'(t)$ for the neural network prediction formula and the linear autoregressive prediction formula. They are input to a neural network and a linear autoregressive prediction formula, and the output function $f^{NN}(\boldsymbol{x}(t))$ is multiplied by a coefficient α. And $f^{AR}(\boldsymbol{x}'(t))$ is multiplied by $(1-\alpha)$. The coupling coefficient α is assumed to have a real value in the range 0 to 1.

$$\begin{aligned} \boldsymbol{x}(t) = [&E_{SA}(t), E_{SA}(t-1), E_{SA}(t-2), P(t), P(t-1), \\ &P(t-2), L(t+1), L(t), T_O(t), T_O(t-1), T_O(t-2)]^T \end{aligned} \tag{6.3}$$

$$y^{NN}(\boldsymbol{x}(t)) = Sigmoid\left(\sum_{j=1}^{J} w_j \times Sigmoid\left(\sum_{i=1}^{I} u_{ij} \times \boldsymbol{x}_i(t)\right)\right) \tag{6.4}$$

$$f^{NN}(\boldsymbol{x}(t)) = y^{NN}(\boldsymbol{x}(t)) \tag{6.5}$$

In neural network prediction equation, input state variable vector $\boldsymbol{x}(t)$ can be obtained using Eq. 6.3, and $f^{NN}(\boldsymbol{x}(t))$ using Eqs. 6.4 and 6.5.

Here, the neural network prediction model perceptron type, and $Sigmoid(z) = 1/\left(1+e^{-az}\right)$ is a sigmoid function in Eq. 6.5. i (=1, 2, …) is input layer node number, I (= 11) is the total number of nodes in the input layer, j (= 1, 2, …) is the node number in the middle layer, J (= 15) is the total number of nodes in the middle layer. u_{ij} is the connection weight from the input layer to the intermediate layer, and w_j is the connection weight from the intermediate layer to the output layer.

This linear autoregressive prediction formula predicts the room temperature index E_{SA} $(t+1)$ one minute after the present time t. E_{SA} $(t+5)$ after 5 min is predicted by repeating the linear autoregressive calculation 5 times. The specific linear autoregressive prediction equation is assumed that the input state variable vector $x(t)$ is given by Eq. 6.6 and $f^{NN}(\boldsymbol{x}(t))$ is given by Eqs. (6.7–6.9).

$$\boldsymbol{x}'(t) = [E_{SA}(t), P(t), L(t)]^{\mathrm{T}} \tag{6.6}$$

$$\mathbf{y}^{AR}(t) = [y_1^{AR}(t), y_2^{AR}(t)]^{\mathrm{T}} = [E_{SA}(t), P(t)]^{\mathrm{T}} \tag{6.7}$$

$$\mathbf{y}^{AR}(t+1) = \sum_{l=1}^{M} \mathbf{A}_l \cdot \mathbf{x}'(t-l) + E_P(t) \tag{6.8}$$

$$f^{AR}(\boldsymbol{x}'(t)) = y_1^{AR}(t + t_F) \tag{6.9}$$

where $\boldsymbol{y}^{AR}(t+1)$ is the predicted state variable vector one step after the discrete time t, $\mathbf{x}'(t-l)$ is the input state variable vector before l steps, and A_l, corresponds to l, is a linear autoregressive coefficient matrix. The elements of the state variable vector consist of power suppression command $L(t)$, the actual power consumption $P(t)$, and the weighted average room temperature index $E_{SA}(t)$. The order M of the linear autoregressive prediction formula is $M = 7$ using the **AIC (Akaike Information Criterion)** method.

Following shows an example of developing a prediction model for the average temperature deviation after 5 min in response to a power restriction command for building multi-type air-conditioning based on specific actual data. Table 6.1 shows the outline of the building multi-type air-conditioning equipment used as an example. This building is a general small office building, and has an average heat load characteristic with an energy consumption rate almost equal to the domestic average. Multi-type air-conditioners for buildings are also standard mass-produced products with typical energy efficiency.

Table 6.1 Example of experimental conditions for building multi air-conditioner FastADR room temperature variations prediction formula

Item	Specifications
Building type	General office building
Oldness	About 20 years
Size	2 story, 1600 m^2
Building structure	Steel concrete
Number of outdoor units	Data collected from 3 out of 5
Number of indoor units	Data collected from 18 out of 30
Rated cooling capacity	40, 45, 68 kW
Energy efficiency (COP)	3.43, 3.04, 3.19

The time series data of air-conditioning power consumption $P(t)$, average room temperature deviation $E_{SA}(t)$, and FastADR power suppression command $L(t)$ are obtained from the actual air-conditioning equipment including 18 indoor units and 3 outdoor units. The interval of time series data is 1 min. The power suppression $L(t)$ was carried out many times for step response tests every 10 min, and learning data for predictive formula identification was collected. The collection period was 11 days in summer, and the maximum temperature during that period was 27.5–33.1 °C.

The collected actual time-series data set was 340 FastADR step responses, and the effective data set was 340 sets except for the complete stop state with zero power consumption. These observed data sets were divided into three groups, namely learning, optimization, and evaluation, and divided into 166, 78, and 96 sets, respectively.

The FastADR room temperature variation prediction formula for predicting the mean room temperature deviation $E_{SA}(t + 5)$ 5 min after the start of FastADR was used for learning. Each variable of the state variable vector was normalized in the range of 0.0 to 1.0 and trained using each training data set 100,000 times. Linear autoregression was trained by the least square method using the same training data set. The weighting coefficient α for combined prediction of neural network with linear autoregressive prediction formula is used for searching the 78 datasets which are not used in learning and testing of the prediction formula to determine the optimal value.

Figure 6.12 shows an example of comparison with actual measurement using a prediction formula. As indicated by the arrow at 15:30, the slope of the room temperature index $E_{SA}(t)$ changes after the power suppression $L(t)$ is activated. The difference between the predicted $E^*_{SA}(t+5)$ and the actual $E_{SA}(t+5)$ was measured, and the performance of the prediction formula was evaluated using *RMSE* as a

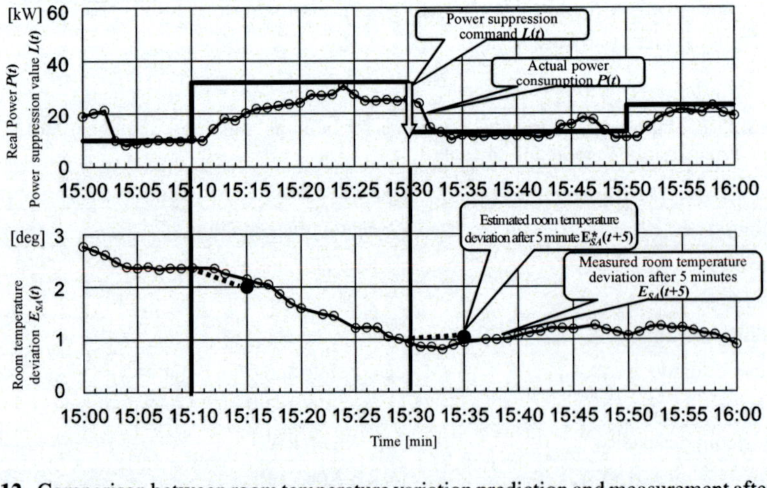

Fig. 6.12 Comparison between room temperature variation prediction and measurement after 5 min for FastADR power suppression by complex neural network prediction formula

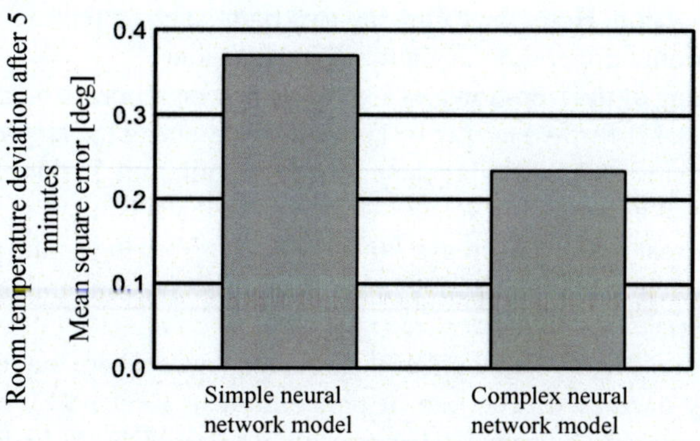

Fig. 6.13 Comparison of root mean square error (RSME) of simple and complex neural networks that predict room temperature variations after 5 min due to FastADR power suppression [6]

statistical method.

$$RMSE = \sqrt{\left\{ \sum_{d=1}^{D} \left(E_{SA}^{*(d)}(t+5) - E_{SA}^{(d)}(t+5) \right)^2 \right\} \Big/ D} \tag{6.10}$$

D is the total number of test data sets, and d (= 1, 2, ...) is the power suppression step response experiment number.

The best FastADR room temperature variation prediction formula based on the optimized data was obtained with $\alpha = 0.53$. Figure 6.13 compares the root mean square error RMSE of the simple neural network prediction formula and the complex neural network prediction formula. The RMSE of each prediction formula for the simple neural network and complex neural network was 0.37 °C and 0.23 °C, respectively. This is an example in which the complex neural network prediction formula was superior to the RMSE of simple neural network prediction formula by about 40%.

The actual prediction performance depends on the individual case, but we will develop a neural network prediction formula that grasps the room temperature variation due to FastADR power suppression for building multi-type air-conditioner.

6.2.2 Prediction Example for Room Temperature Variation Due to FastADR

This section introduces an example of evaluating the effect of FastADR power suppression command $L(t)$ on the average room temperature deviation $E_{SA}(t)$. This evaluation is based on the composite neural network prediction formula described in previous subsection. Here, the following two points need attention when evaluating the average room temperature deviation $E_{SA}(t)$ response.

First, looking at the step-down of FastADR power suppression command $L(t)$ at 15:30 in Fig. 6.12. $P(t-1) < L(t-1)$ one-minute-ahead the step-down, the actual power consumption was not using the power to the full limit. In this case, the effective power suppression amount is not $L(t-1) - L(t)$, but it is more meaningful for the evaluation to treat it as the decrease $P(t-1) - L(t)$ from the actual power. Here, the amount of reduction in the effective power suppression command is defined as $\Delta L_E = |P(t\text{-}1) - L(t)|$.

Another point to note is that under the circumstances where the room temperature has originally decreased. The rate of temperature is reduced by reducing the air-conditioning capacity by suppressing power with FastADR. At first glance, it seems that the room temperature rises when FastADR power is suppressed. However, the room temperature may not increase due to the heat balance, and even if there is a delay, the room temperature starts to rise after 5 min.

For example, if the air-conditioning power is suppressed with FastADR as shown in 15:30 in Fig. 6.12, the average room temperature deviation of the entire building may increase after 5 min. However, in this situation where $E_{SA}(t)$ is in a downward trend, that is, in a situation where the air-conditioning capacity exceeds the heat load, the downward trend is alleviated but does not turn upward.

Therefore, the influence of FastADR power suppression $L(t)$ after 5 min must evaluate quantitatively the change in average room temperature deviation trend. And not the average room temperature deviation $E_{SA}(t+5)$ value itself after 5 min. Here, the trend of average room temperature deviation $E_{SA}(t+5)$ across the building before and after the start of FastADR is defined as $\Delta G_{SA}(t+5)$ by Eq. 6.12.

$$G_{SA}(t+5) = E_{SA}(t+5) - E_{SA}(t) \tag{6.11}$$

$$\Delta G_{SA}(t+5) = G_{SA}(t+5) - G_{SA}(t) \tag{6.12}$$

In this way, the FastADR power suppression command $L(t)$ is distributed to each building as viewed from the FastADR aggregator. The amount of power reduction from the actual power $P(t)$ is predicted by prediction formula immediately before the distribution of effective power suppression command ΔL_E. Also, $E_{SA}(t+5)$ after 5 min is calculated using the prediction formula, and the comfort side is quantitatively evaluated from the trend change $G_{SA}(t+5)$ of the temperature index.

Figure 6.14 is a plot of trend change $\Delta G_{SA}(t+5)$ of mean room temperature deviation against effective power suppression command reduction width ΔL_E in 174 FastADR tests of an actual building. These 174 times include 78 data sets used to optimize FastADR room temperature change prediction coefficient α.

The rated power consumption P_R of all air-conditioning equipment for this office building is 51 kW. So, for example, the reduction of $\Delta L_E = -10$ kW corresponds to 20%. Figure 6.14a shows the predicted values based on our linear auto-regressive neural network FastADR room temperature change prediction formula.

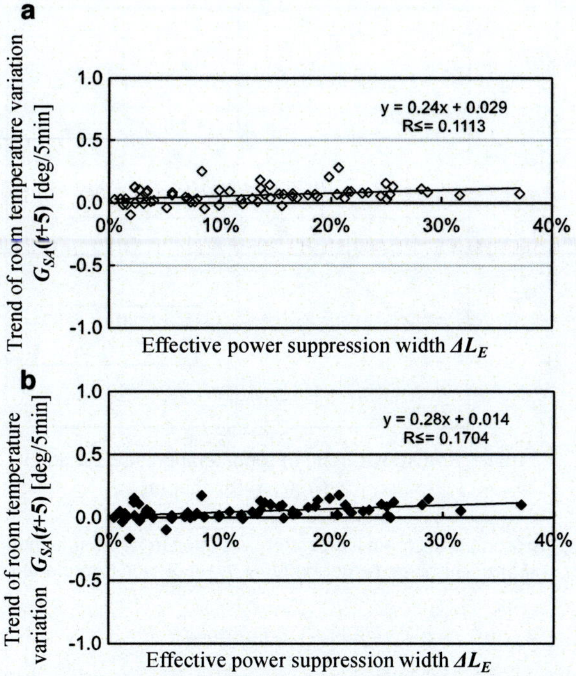

Fig. 6.14 Prediction experiment on the influence of the building multi air-conditioner FastADR power on trend after 5 min of the average temperature deviation of the entire building [6]

And Fig. 6.14b shows the measured values that can be compared with each predicted value.

If ΔL_E is reduced by 0–10 kW (relative reduction of 0–20%), both prediction (a) and actual measurement (b) are buried in the original air-conditioning power and stochastic fluctuations in room temperature. $G_{SA}(t+5)$ is not always positive. In other words, the trend of average room temperature deviation of the entire building did not always worsen.

From the region where the decrease in ΔL_E exceeds approximately − 10 kW (relative decrease 20%), both the prediction (a) and the actual measurement (b) gradually decrease and all $G_{SA}(t+5)$ are positive. The average room temperature deviation trend all worsened. As can be seen from the measurement example in Fig. 6.12, the power $P(t)$ and the mean room temperature deviation $E_{SA}(t)$ are almost quantitatively consistent with stochastic fluctuation.

The degree of influence of the FastADR power suppression was plotted as the slope of the regression line of $\Delta G_{SA}(t+5)$ versus ΔL_E in Fig. 6.14a and (b). As a result, the regression line was $\Delta G_{SA}(t+5) = 0.24\Delta L_E + 0.029$ in the prediction of Fig. 6.14a, and the slope was 0.24 per 10% ΔL_E.

In Fig. 6.14b, the regression line was $\Delta G_{SA}(t+5) = 0.28\Delta L_E + 0.014$, and the slope was 0.28. Both agreed with being about a quarter of 1 °C. This means that even

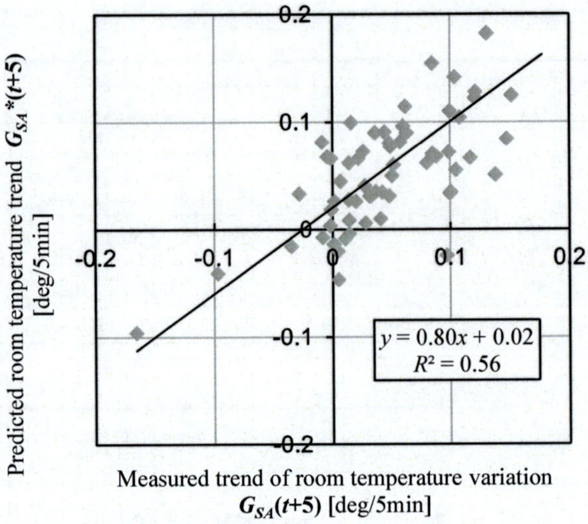

Fig. 6.15 Example of comparing room temperature variation trend of entire building by FastADR to building multi air-conditioner in actual office building by prediction formula and actual measurement [6]

if the reduction rate ΔL_E of the power suppression command is changed, the average room temperature deviation trend change $\Delta G_{SA}(t+5)$ is 5% immediately after that. And the reduction rate is 1/4 of 1 °C per 10% of the rated power consumption.

Next, Fig. 6.15 shows the correspondence between the predicted $\Delta G_{SA}{}^{*}(t+5)$ and the measured $\Delta G_{SA}(t+5)$ for each temperature index trend. This plot is plotted with the horizontal axis of the same test point as the actual value and the vertical axis as the predicted value. The closer the slope of each point is to the straight line of 1, the closer the predicted value and the measured value match.

As a result, the regression line $y = 0.80x + 0.02$ of all points had a slope of 0.8, which was relatively close to the slope 1. The correlation coefficient for the regression line was $R = \sqrt{0.56}$, and $R > 0.7$, which is said to be a strong correlation.

Although this time is an example of one office building and lacks generality, this example is a total value of three mass-produced types of general multi-type air-conditioning units for buildings in a very small office building.

For a common multi-type air-conditioning system for buildings, we experimented 174 times with the FastADR under various conditions. And we considered what the result of comparing with the prediction formula means as follows.

As shown in Fig. 6.14, when the reduction range of effective power suppression is in the range of 0 to 40% of rated power consumption, the average room temperature deviation trend change $\Delta G_{SA}(t+5)$ after 5 min is the effective power suppression reduction range. It was found that it was hardly affected by ΔL_E. Quantitatively, the correlation of the regression line shown in Fig. 6.14 is $R^2 = 0.11$ in measured valu e and $R^2 = 0.17$ in predicted value. And both the measured value and predicted value are almost the same, with little correlation.

Also, according to the slope of the regression line the bend of room temperature index trend line is about 0.24 °C/5 min in prediction formula, and actual measurement affects only about 0.28 °C/5 min. Assuming that the range that does not affect most residents initially assumed in this chapter is within 0.5 °C, and the FastADR sustainable time is considered to be 9–10 min.

The reason is that in this building air-conditioning system, the change in the average room temperature of the entire building is delayed as a result after 5 min, regardless of the ΔL_E reduction, and the change has not yet appeared.

In fact, this is supported by the fact that the room temperature change time constant was about 15 min due to the step response of all air-conditioning facilities at the time of lunch break or leaving office. In other words, many literature described that "even if the air-conditioning power is suppressed for a short time, the rise in room temperature of the duilding does not change so much that the residents feel uncomfortable". Although this is a typical building, it seems to be an assessable example.

So far, FastADR studies using air-conditioning have not been assessable addressed the consequences of room temperature rise, that is, the comfort of residents due to power suppression. There is a possibility that the power suppression command can be distributed to each building if the FastADR power suppression command to the air-conditioning equipment is at least about 5 min while maintaining the comfort level of each building within the predetermined range.

Since the FastADR duration of 5 min or more is necessary for the entire power system. The aggregator should rotate the target building air-conditioning equipment in units of about 5 min or apply feedback control in about 5 min.

The average room temperature deviation $E_{SA}(t)$ imposes a constraint that gives one index for the entire building. Of course, since the temperature change allowable range for each indoor unit in the building is different. The power command in the building will be further subdivided and adjusted by local control of each building. So that the individual room temperature deviation is traced within the range.

However, from the aggregator's point of view, it is necessary to issue a distribution command to each building within a short period of time, for example within one minute. And the room temperature index for each building will be list up for a practical management.

This time, the prediction formula was developed from the actual measurement data of the building. But in practical application, it would be necessary to repeat research to develop several types of standard prediction formulas by accumulating the results of the prediction formula.

6.3 Negawatt Aggregation of Building Multi-type Air-Conditioners

6.3.1 Averaging Effect of Building Multi-type Air-Conditioning Facility

Since solar power generation and wind power generation generate electricity by natural phenomena, the amount of power generated varies uncertainly. However, it is known that the individual variations are smoothed by the averaging effect of aggregating them in large quantities.

As for building multi-type air-conditioning, preservation of built-in control for refrigerant circuit, equipment protection, etc. must be performed on the spot. So individual air-conditioners could not always follow the power suppression command value. In other words, as shown in Fig. 6.16, the fractional power response waveform of each building multi-type air-conditioner has a time period that exceeds the power suppression command value transiently, and conversely, it does not consume the power suppression command value. There is a band, and the response waveform is indeterminate. However, it has been stated that if these uncertain responses are aggregated in large quantities, the individual behavior is smoothed by the averaging effect. And the integrated power after aggregation converges to a certain expected value [10].

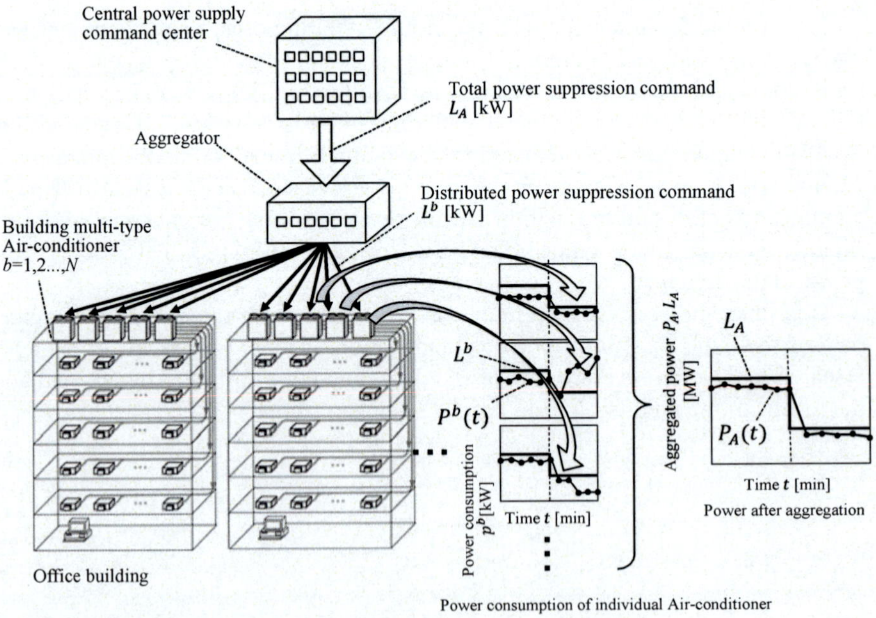

Fig. 6.16 FastADR aggregation of building multi-type air-conditioning

The difference between the post-aggregation power suppression target value L_A[kW] and the post-aggregation power $P_A(t)$[kW] evaluated for 5 min is referred to as FastADR margin R_M[%] in this book. The following formula is defined the FastADR margin R_M[%].

$$R_M = \left(\sum_{t=1}^{\tau_m} P_A(t) - \sum_{t=1}^{\tau_m} L_A \right) \Bigg/ \sum_{t=1}^{\tau_m} L_A \tag{6.13}$$

$$P_A(t) = \sum_{b=1}^{N} P^b(t) \tag{6.14}$$

$$L_A = \sum_{b=1}^{N} L^b \tag{6.15}$$

Here, $P^b(t)$ is the power consumption [kW] of air-conditioner b, L^b is the power suppression command value [kW] for air-conditioner b, and τ_m is the interval [min] of the power suppression command value.

Figure 6.17 shows the relationship between the distribution of FastADR margin R_M (average value $\mu \pm$ standard deviation σ) and the number N of aggregates when the power suppression command value in downward direction is an aggregated. As shown in the Fig. 6.17, when the number of aggregation N is small, the dispersion of R_M is large. On the other hand, as N increases, the standard deviation σ of R_M decreases, and eventually converges to the mean value μ of distribution and remains as an offset. At this time, the rate of decrease of the standard deviation σ is $1/\sqrt{N}$ times.

Well, since the power suppression command is a simple open loop control, an offset occurs. Although it may be improved if closed-loop control is applied. But stability is a problem depending on the design of the closed-loop controller because

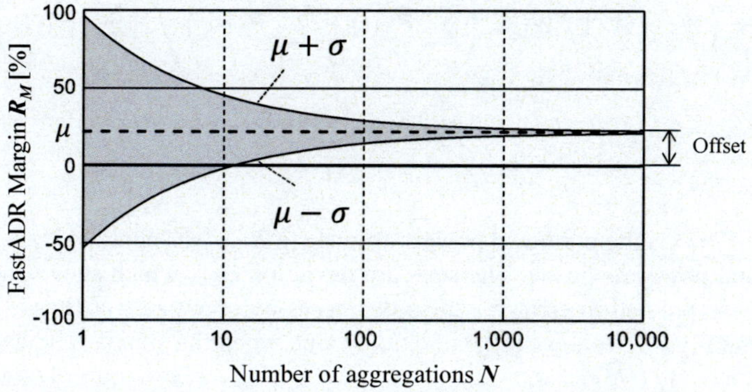

Fig. 6.17 Relationship between number of aggregations and FastADR margin

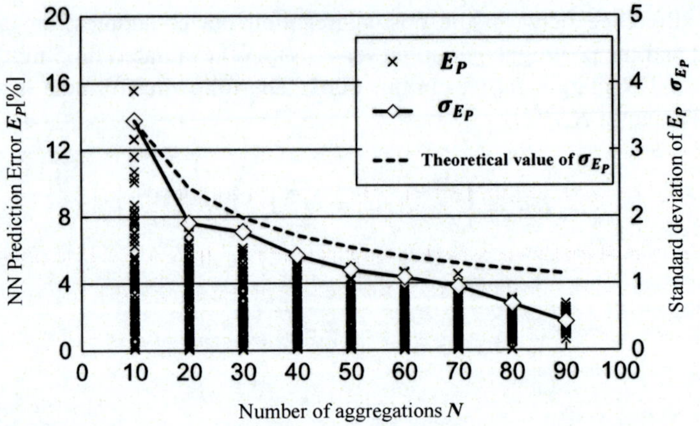

Fig. 6.18 Relationship between the number of aggregations and FastADR power suppression response prediction error

FastADR closed-loop control causes dead time such as wide area communication delay and FastADR procedure [11–15].

On the other hand, if the aggregated power response value in large aggregation can be quantitatively predicted, the required accuracy in practical use may be satisfied even in open-loop control. However, to do this, the uncertain behavior of air-conditioning aggregated power must be averaged and mitigated.

On the other hand, the power prediction by the neural network described earlier is suitable for simulating nonlinear uncertain behavior. Of course, a certain amount of error occurs between the predicted value and the measured value. But as shown in Fig. 6.18, the response prediction error E_P[%] of the aggregated 5-min power decreases as the N number of aggregations increases. The definition of E_P is as Eq. 6.16.

$$E_P = \left(\tilde{W}_5 - W_5\right) \Big/ W_5 = \left(\sum_{t=1}^{5} \widetilde{P_A}(t) - \sum_{t=1}^{5} P_A(t)\right) \Big/ \left(\sum_{t=1}^{5} P_A(t)\right) \tag{6.16}$$

$$\widetilde{P_A}(t) = \sum_{b=1}^{N} \widetilde{P^b}(t) \tag{6.17}$$

Here $\widetilde{P^b}(t)$ is the predicted power response [kW] of air-conditioner b, and $\widetilde{P_A}(t)$ is the total power response. The standard deviation σ_{E_P}, which shows the variation in response prediction error E_P, also decreases by about $1/\sqrt{N}$ times. The dotted line in the Fig. 6.18 shows the theoretical value when the standard deviation σ_{E_P} is decreased by $1/\sqrt{N}$ times.

In this example, the avergaing effect is noticeable even at relatively small N, and simply increasing from $N = 10$ to $N = 50$ reduces the standard deviation around 50%. Furthermore, the response prediction error E_P at $N = 90$ is only about 2%.

In this way, for prediction of large aggregation response, the prediction error of the aggregation power can be expected to be reduced by canceling the prediction errors of the individual power responses.

Thus, due to follow the central limit theorem, variation of FastADR margin R_M and neural network response prediction error E_P decreases by $1/\sqrt{N}$. **Central limit theorem** means that a random variable x with a population mean of μ and a population variance of σ^2 has a sample sum $\sum_{n=1}^{N} X_n$ or when N of the sample mean $\left(\sum_{n=1}^{N} X_n\right)/N$ becomes larger. It means that it approaches the normal distribution. This is shown as Eq. 6.18 after variable conversion to a standard normal distribution.

$$\lim_{N\to\infty} \Pr\left\{x_a \le \left(\sum_{n=1}^{N} X_n - N\mu\right) \Big/ \left(\sqrt{N}\sigma\right) \le x_b\right\}$$
$$= \Phi(x_b) - \Phi(x_a) = \int_{x_a}^{x_b} \frac{1}{\sqrt{2\pi}} \exp\left[-\frac{x^2}{2}\right] dx \tag{6.18}$$

Here, x_a and x_b are the realization value specification ranges, and $\Phi(x)$ is the cumulative distribution function of the standard normal distribution N (0,1).

This is applied to the nth iteration real value $W_m^{(n)}$ of FastADR power suppression in the mth building. Even if each actual value is not a normal distribution, its repeated sample sum W_{AG} means that as N increases. It approaches the normal distribution, that is, even if there is variation, it means that the number of air-conditioning units will be increased.

Now, when the iterative real value is calculated with the predicted value $W_m'^{(n)}$, the normal distribution with the sample sum average of $N \cdot \mu_W$ and sample sum variance of $N \cdot \sigma_W^2$ is obtained. Where μ_W and σ_W^2 are the mean and variance of the original distribution of $W_m^{(n)}$.

$$W_{AG} = \sum_{m=1}^{M} W'^{(n)}_m \tag{6.19}$$

$$\lim_{N\to\infty} \Pr\left\{ W_{AGa} \le \sum_{m=1}^{M} W'^{(n)}_m \le W_{AGb} \right\}$$
$$= \int_{W_{AGa}}^{W_{AGb}} \frac{1}{\sqrt{2\pi}\sqrt{N}\sigma_W} \times \exp\left[-\frac{(W_{AG} - N\mu_W)^2}{2N\sigma_W^2}\right] dW_{AG} \tag{6.20}$$

Here, the central limit theorem holds for the iterative sample means. The iterative average is itself, and the variance of the iterative sample average is close to a normal distribution. Also, in FastADR margin R_M, the variance reduction rate approaches $1/\sqrt{N}$, which can be said to be applied to the central limit theorem.

Note that the conditions from the **ergodic stochastic process** $x^{(k)}(t)$ is that k is an individual number, vector $\boldsymbol{x}(t)$ is an individual set, $g()$ is an arbitrary function, and $\boldsymbol{E}[]$ is a combined average. It is expressed by the Eq. 6.21.

$$\lim_{T\to\infty}\frac{1}{T}\int_0^T g\left(x^{(k)}(t), x^{(k)}(t+\tau_1), \cdots, x^{(k)}(t+\tau_n)\right)dt$$
$$= \boldsymbol{E}[g(\boldsymbol{x}(t), \boldsymbol{x}(t+\tau_1), \cdots, \boldsymbol{x}(t+\tau_n))] \tag{6.21}$$

6.3.2 *Comfort Preservation and Rotational Control*

The basic demand response in conventional building multi-type air-conditioning facility was based on a simple method of selecting an air-conditioner that would not be affected even if it was stopped in advance and stopping it in response to the demand response. In this case, since the demand response duration is as long as 30 min or 1 h, the consequences that the room temperature always rises or falls will occur. Therefore, in such a conventional method, it is necessary to set the target air-conditioner for each facility and each date according to the usage situation.

However, if FastADR technology that suppresses power in minutes is realized, the above situation may be solved at once. In FastADR, the power suppression command can be adjusted in increments of minutes. So it is technically possible to control the FastADR power suppression command with a small width of about 20–30% compared to the rated power consumption for a few minutes. With such a small and short-time suppression, it is possible to maintain feeling of resident into comfort zone.

This section introduces the FastADR packetized rotation method as an idea to solve the room temperature change trade-off during building air-conditioning FastADR. This means that the FastADR power suppression command value and duration for each air-conditioner are subdivided into "demand response packets". And the group dynamically performs "rotation" so that the virtual power plant as a whole follows the power feed command.

Here, as an example, FastADR for each air-conditioner is standardized into fine suppression units and short duration units, and FastADR that suppresses $\Delta L = 1$ kW for 5 min is called "1 packet". Then we simulate an example of generating 1000 W $\times$ 3 groups = 3000 W of 5 MW $\times$ 30-min duration of power suppression as a virtual power plant.

Each air-conditioner simulates a total of 30 min of operation in which 3 groups rotate twice, with FastADR of 5 packets (inhibition command $\Delta L = 5$ kW $\times$ 5 min) for each air-conditioner, with a time difference of 5 min.

It is also assumed that each air-conditioner was in steady stable operation at 26 °C at the start of FastADR rotation with a power consumption P of 15 kW and both the set room temperature T_{SET} and the measured room temperature T_{AIR}. In this scenario, each group shifts 5 packets by 5 min from the initial state. It was assumed that the response time delay of each group occurred uniformly from 0 to 5 min. In addition, when the packet time is 5 min for each group, the power is limited to the starting power (15 kW in this initial setting) for the next 10 min, and then the rotation is continued or the demand response is terminated.

Figure 6.19 shows the simulation results. Figure 6.19a shows the simulation results of the change in power and room temperature that each building multi-type-air-conditioner executed two rotations. As each air-conditioner only suppresses power for 5 min per rotation, it is assumed that the room temperature rise is suppressed to 0.5 °Cfor the first time and within 1 °C for the second time as shown in Fig. 6.19b.

Here, assuming that the demand response duration is 30 min for the entire virtual power plant, 3 groups will execute 2 rotations each. Figure 6.19c shows the result of aggregation simulation of the total power suppression in this case.

In this way, each air-conditioner only suppresses power for 5 min, which is one FastADR packet time, but the aggregate power suppression amount for the virtual power plant as a whole is the required suppression amount $\Delta P = -5$ MW can be secured continuously.

After end of the overall demand response duration, the power consumption of each air-conditioner increases transiently from the start of the overall demand response and rebound occurs, trying to return to the temperature change. This is because, in this simulation experiment, the end of rotation is the same for all air-conditioners in the group. In other words, by adjusting the time until the power suppression command value is completely released within the group, it is possible to smooth the rebound after the end of the overall demand response as shown in Fig. 6.19d. By repeating such a simulation study, the optimal FastADR packetization rotation method will be realized.

6.4 Compensation for Building Multi-type Air-Conditioner Facility FastADR Using Storage Battery

6.4.1 Building Multi-type Air-Conditioning FastADR Followability to Supply and Demand Adjustment Signal

In the previous section, it was shown that uncertainties decrease due to the averaging effect when aggregation is performed in large quantities. However, even if the uncertainty is smoothed and the fluctuation is reduced, the building multi-type air-conditioner does not completely match the FastADR power suppression command value. Even if the resource aggregator distributes the FastADR power suppression

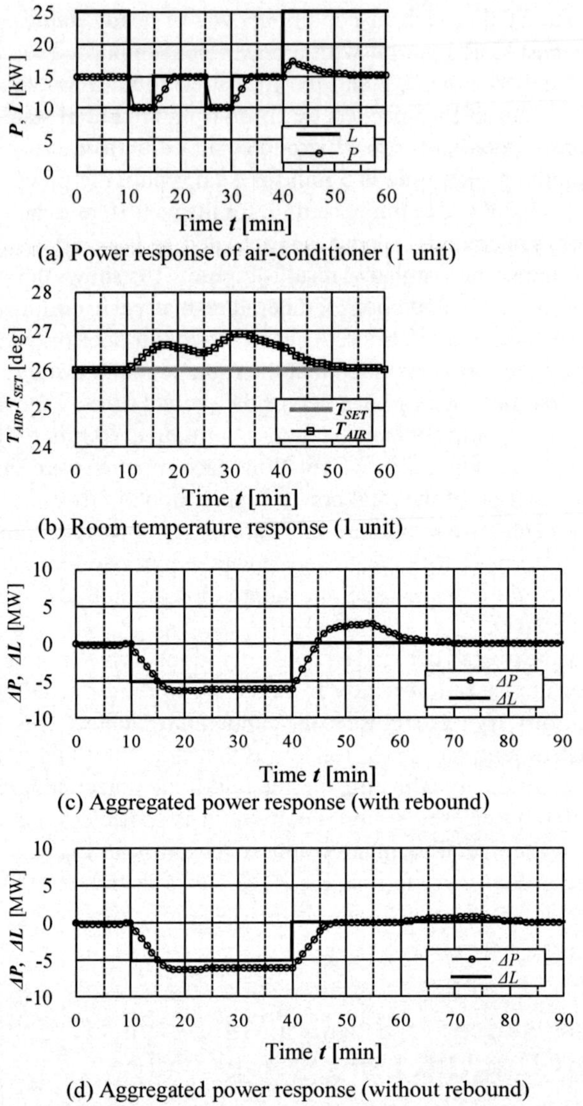

Fig. 6.19 FastADR packetization rotation method and rebound suppression

command to each building multi-type air-conditioner. It is not perfect even if high-speed feedback control is performed on each target air-conditioner response power. It is impractical to expect to match the FastADR command signal waveform completely in the transient state of suppression during start and release period, as well as when FastADR control is stable.

The first reason is the complex built-in control of the refrigerant circuit, which is given priority over the physical characteristics of the original building multi-type air-conditioner, and then to approaching the control command value. Furthermore, because of the transient power increase/decrease speed due to the acceleration/deceleration of the inverter compressor that determines the actual power waveform, the power waveform with reduced power is reduced in response to a step change in the command value or a sudden lamp state change.

Therefore, it can be considered to compensate for the tracking error that is inevitable when generating negawatts with such a building multi-type air-conditioner by charging and discharging the storage battery system. This method has not been fully studied and no published literature is available. In this section, we explain the dynamic characteristics of the building multi-type air-conditioner FastADR power suppression introduced in Chap. 7 and the basic simulation using the model created from the building heat balance.

The charge/discharge model of the storage battery system used here simulates the SOC of the storage battery system as the simplest and not discussed in the book.

Assuming that each power is reported at a period of $\tau_{BC} = 30$ s, the aggregated power $P_A(t)$ is calculated by summing up the total power for the number of target air-conditioners. Here, time t is the discrete time of the control cycle $\tau_{BC} = 30$ s of the storage battery control system. The response aggregated power value $P_A(t)$ and the FastADR aggregate command value $L_A(t)$ are compared every 30 s, and a control signal for charging or discharging is issued to follow-up error. The charge/discharge power is commanded from the resource aggregator to the controller of the storage battery system. Then, $P_{BAT}(t)$ is generated as positive watts or negative watts, and added to the building multi-type air-conditioner aggregated power $P_A(t)$. Since the storage battery system compensates to match the follow-up of the building multi-type air-conditioning aggregated power, this combined power is (provisionally) named compensate ion power $P_{COM}(t)$.

Figure 6.20 shows an example of FastADR aggregated power response simulation for 100 building multi-type air-conditioners without storage battery compensation. At time $t = 0$, the FastADR power suppression command $L(t)$ indicated by the dotted

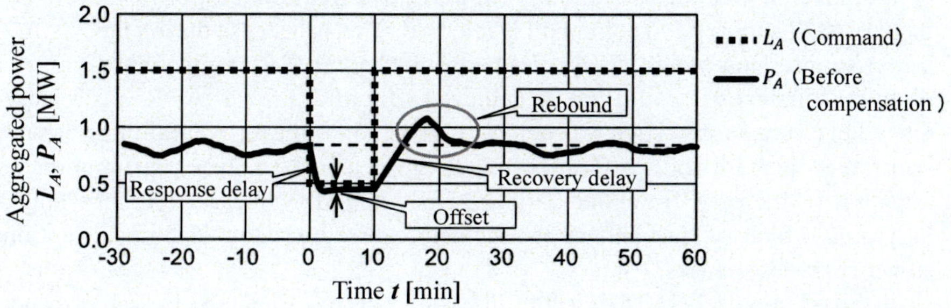

Fig. 6.20 Building multi-type air-conditioning group FastADR response power follow-up response to supply/demand adjustment signal

line is activated, and a target value (dotted line) is given to reduce the aggregate power $P_A(t)$ to 0.5 MW.

In response to this, the building multi-type air-conditioner unit deaccelerates the inverter compressor speed, but the response delay is unavoidable due to the original characteristics of the air-conditioner. Since the speed of inverter compressor stables in less than 1 min, it can be completed in a relatively short time.

In addition, in the suppression stable state after completion of deceleration, the power of each air-conditioner may fluctuate uncertainly, but the total of 100 units is considered, stable aggregated power is maintained due to the averaging effect. In this example, the power suppression state is reached, but the aggregate power is slightly lower $P_A(t) = 0.45$ MW than compared to the dotted line command value $L(t) = 0.5$ MW. Since the power consumption responses of 100 air-conditioners are consolidated, uncertain fluctuations are suppressed by the averaging effect, and the power is maintained by suppressed to the stable linear shape. However, how much the offset appears on the plus or minus side varies depending on the air -conditioning model, thermal load conditions, and the degree of demand response suppression.

An easy-to-understandable example, the aggregated power value $P_A(t)$ is suppressed in situations where some air-conditioners fail to follow the FastADR power suppression command and drop out (called OptOut) during the FastADR power suppression time. Since it deviates from the command value $L(t)$, it is inevitable that any offset will occur. In this simulation example, OptOut is not used, and a large amount of aggregation of 100 units is aggregated.

After 10 min, the FastADR power suppression command is released stepwise from $t = 10$. The built-in control of the building multi-type air-conditioner receives the release command, and the inverter compressor accelerates and the power consumption increases. When increasing the compressor rotation speed and increasing the amount of refrigerant circulation. The rate of increase is suppressed due to the restriction of air-conditioner refrigerant control. Therefore, even if the FastADR is released stepwise as shown by the dotted line, the aggregated power value of 100 units consumed by the actual air-conditioner will only increase gradually. In this example, it took 5 min to return to the initial power value at $t = 15$ just as similar as before FastADR.

Moreover, a phnomenon called "**rebound after demand response**" occurs after the FastADR suppression command is released. This is because during the FastADR power suppression period, the cooling capacity required for temperature control at room set temperature could not be achieved. But after the FastADR was released, the temperature control state was being restored. Since the air-conditioning capacity will rather increase for a certain period after release, the power consumption will temporarily increase from the initial value. Naturally, it can occur not only in FastADR, which is a fast and short-time demand response, but also in conventional slow demand responses.

In this example, about 15 min after FastADR was released, rebound settled around $t = 25$, and aggregated power $P_A(t)$ returned to the initial power value. The rebound phenomenon is an unacceptable phenomenon in terms of supply–demand adjustment on the grid side. If the demand response time is long and slow, the demand response may be allowed to be rebounded to some extent immediately after the demand response is released. However, if the demand response time is short, such as 10 min for FastADR of a virtual power plant, the rebound immediately after release continues for several tens of minutes makes FastADR meaningless. Therefore, it is assumed here that after releasing FastADR for 10 min, it is required to limit the power consumption during rebound period at the initial power consumption so as not to rebound for 20 min.

In this simulation, we will try to compensate for the tracking error with the suppression command value due to the above-mentioned response delay, offset, return delay, rebound, etc. by controlling the charge/ ischarge of the power storage battery at high speed.

6.5 Compensation for Building Multi-type Air-Conditioner Facility FastADR Using Storage Battery

6.5.1 Building Multi-type Air-Conditioning FastADR Followability to Supply and Demand Adjustment Signal

In the previous section, it was shown that uncertainties decrease due to the averaging effect when aggregation is performed in large quantities. However, even if the uncertainty is averaged and the fluctuation is reduced, the building multi-type air-conditioner does not completely match the FastADR power suppression command value. Even if the resource aggregator distributes the FastADR suppression command to each building multi-type air-conditioner. It is not perfect even if high-speed feedback control is performed on each target air-conditioner response power. It is impractical to expect to match the FastADR command signal waveform completely in the transient state of suppression during start and release period, as well as when FastADR control is stable.

The first reason is the complex built-in control of the refrigerant circuit, which is given priority over the physical characteristics of the original building multi-type air-conditioner. And then to approaching the control command value. Furthermore, because of the transient power increase/decrease speed due to the acceleration/deceleration of the inverter compressor that determines the actual power waveform, the power waveform with reduced power is reduced in response to a step change in the command value or a sudden lamp state change.

Therefore, it can be considered to compensate for the tracking error that is inevitable when generating negawatts with such a building multi-type air-conditioner by charging and discharging the storage battery system. This method has not been fully studied and no published literature is available. In this section, we explain the dynamic characteristics of the building multi-type air-conditioner FastADR power suppression introduced in Chap. 7 and the basic simulation using the model created from the building heat balance.

The charge/discharge model of the storage battery system used here simulates the SOC of the storage battery system as the simplest and not discussed in the book.

Assuming that each power is reported at a period of $\tau_{BC} = 30$ s, the aggregated power $P_A(t)$ is calculated by summing up the total power for the number of target air-conditioners. Here, time t is the discrete time of the control cycle $\tau_{BC} = 30$ s of the storage battery control system. The response aggregated power value $P_A(t)$ and the FastADR aggregate command value $L_A(t)$ are compared every 30 s, and a control signal for charging or discharging is issued to follow-up error. The charge/discharge power is commanded from the resource aggregator to the controller of the storage battery system. Then, $P_{BAT}(t)$ is generated as positive watts or negative watts, and added to the building multi-type air-conditioner aggregated power $P_A(t)$. Since the storage battery system compensates to match the follow-up of the building multi-type air-conditioning aggregated power, this combined power is (provisionally) named compensated power $P_{COM}(t)$.

Figure 6.20 shows an example of FastADR aggregated power response simulation for 100 building multi-type air-conditioners without storage battery compensation. At time $t = 0$, the FastADR power suppression command $L(t)$ indicated by the broken line is activated, and a target value (dotted line) is given to reduce the aggregate power $P_A(t)$ to 0.5 MW.

In response to this, the building multi-type air-conditioner unit deaccelerates the inverter compressor speed, but the response delay is unavoidable due to the original characteristics of the air-conditioner. Since the speed of inverter compressor stables in less than 1 min, it can be completed in a relatively short time.

In addition, in the suppression stable state after completion of deceleration, the power of each air-conditioner may fluctuate uncertainly, but the total of 100 units is considered, stable aggregated power is maintained due to the averaging effect. In this example, the power suppression state is reached, but the aggregate power is slightly lower $P_A(t) = 0.45$ MW than compared to the broken line command value $L(t) = 0.5$ MW. Since the power consumption responses of 100 air-conditioners are consolidated, uncertain fluctuations are suppressed by the averaging effect, and the power is maintained by suppressed to the stable linear shape. However, how much the offset appears on the plus or minus side varies depending on the air -conditioning model, thermal load conditions, and the degree of demand response suppression.

An easy-to-understandable example, the aggregated power value $P_A(t)$ is suppressed in situations where some air-conditioners fail to follow the FastADR power suppression command and drop out (called OptOut) during the FastADR power suppression time. Since it deviates from the command value $L(t)$, it is inevitable that any offset will occur. In this simulation example, OptOut is not used, and a large amount of aggregation of 100 units is aggregated.

After 10 min, the FastADR power suppression command is released stepwise from $t = 10$. The built-in control of the building multi-type air-conditioner receives the release command, and the inverter compressor accelerates and the power consumption increases. When increasing the compressor rotation speed and increasing the amount of refrigerant circulation. The rate of increase is suppressed due to the restriction of air-conditioner refrigerant control. Therefore, even if the FastADR release command is released stepwise as shown by the dotted line, the aggregated power value of 100 units consumed by the actual air-conditioner will only increase gradually. In this example, it took 5 min to return to the initial power value at $t = 15$ just as similar as before FastADR.

Moreover, a phnomenon called "**rebound after demand response**" occurs after the FastADR suppression command is released. This is because during the FastADR power suppression period, the cooling capacity required for temperature control at room set temperature could not be achieved. But after the FastADR was released, the temperature control state was being restored. Since the air-conditioning capacity will rather increase for a certain period after release, the power consumption will temporarily increase from the initial value. Naturally, it can occur not only in FastADR, which is a fast and short-time demand response, but also in conventional slow demand responses.

In this example, about 15 min after FastADR was released, rebound settled around $t = 25$, and aggregated power $P_A(t)$ returned to the initial power value. The rerbound phenomenon is an unacceptable phenomenon in terms of supply–demand adjustment on the grid side. If the demand response time is long and slow, the demand response may be allowed to be rebounded to some extent immediately after the demand response is released. However, if the demand response time is short, such as 10 min for FastADR of a virtual power plant, the rebound immediately after release continues for several tens of minutes makes FastADR meaningless. Therefore, it is assumed here that after releasing FastADR for 10 min, it is required to limit the power consumption during rebound period at the initial power consumption so as not to rebound for 20 min.

In this simulation, we will try to compensate for the tracking error with the suppression command value due to the above-mentioned response delay, offset, return delay, rebound, etc. by controlling the charge / discharge of the power storage battery at high speed.

6.5.2 *Compensation of Tracking Error by Storage Battery of Building Multi-type Air-Conditioner Facility FastADR*

Figure 6.21 shows a conceptual diagram of a system in which a virtual power plant resource aggregator performs high-speed charge/discharge control of a storage battery system. This charge/discharge control is performed in response to a FastADR response of a building multi-type air-conditioner facility for one building. In this system, the resource aggregator of the virtual power plant is also required to receive the entire negawatt power generation command $L_A(t)$ from the higher-level aggregation coordinator. For simplicity, aggregation coordinator is not shown in this Fig. 6.21.

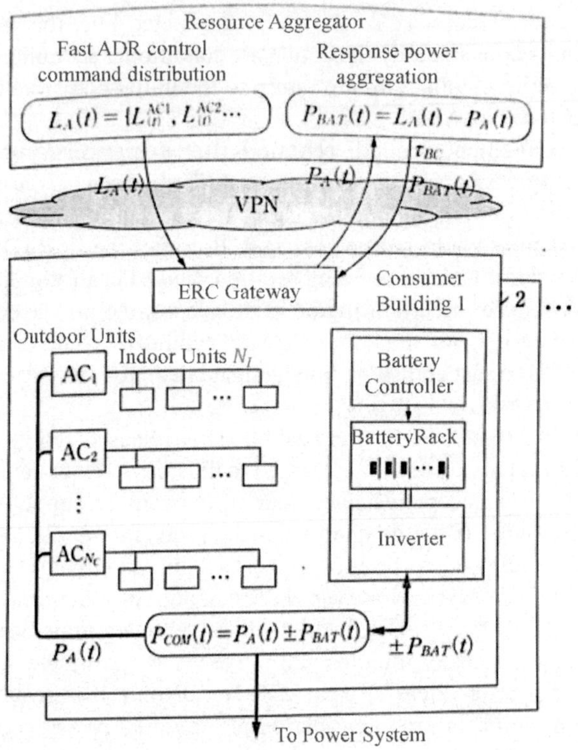

Fig. 6.21 Building multi air-conditioner group FastADR aggregation Virtual power generation follow-up compensation control system for power generation command value by battery system

Based on the required total negawatt power generation command value $L_A(t)$, Resource aggregator distributes to each air-conditioner *b1*, b2, … as FastADR suppression commands $L^{b1}(t)$, $L^{b2}(t)$, … The set of distributed FastADR power suppression commands is notified to the ERC gateway device of customer building side. And each building multi-type air-conditioner *b1*, b2, … respectively suppress the power. The ERC reports the response power to the resource aggregator at every control cycle $\tau_{BC} = 30$ s.

The resource aggregator measures the tracking error by comparing the aggregated response power $P_A(t)$ every $\tau_{BC} = 30$ s with the total suppression command value $L_A(t)$ commanded by itself. In order to compensate for this, the battery charge / discharge value $P_{BAT}(t)$ every $\tau_{BC} = 30$ s is notified to the ERC gateway of the customer building.

In this simulation, the control delay is also controlled in the $P_{BAT}(t)$ output of the storage battery system.

It is assumed that there is no error. In reality, there may be a delay of seconds or an error of several percent or less, but these are ignored in this simulation.

Figure 6.22 shows the result of compensation control simulation for improving the tracking error by the storage battery system for the aggregated power of the building multi-type air-conditioning FastADR. Table 6.2 shows the parameter settings for this simulation. For ease of comparison, Fig. 6.21 is shown again as Fig. 6.22a.

Figure 6.22b shows the result of tracking error compensation by the storage battery for the building multi-type air-conditioning FastADR response power. Compared with Fig. 6.22a (that is, Fig. 6.21), the effect of the storage battery compensation is

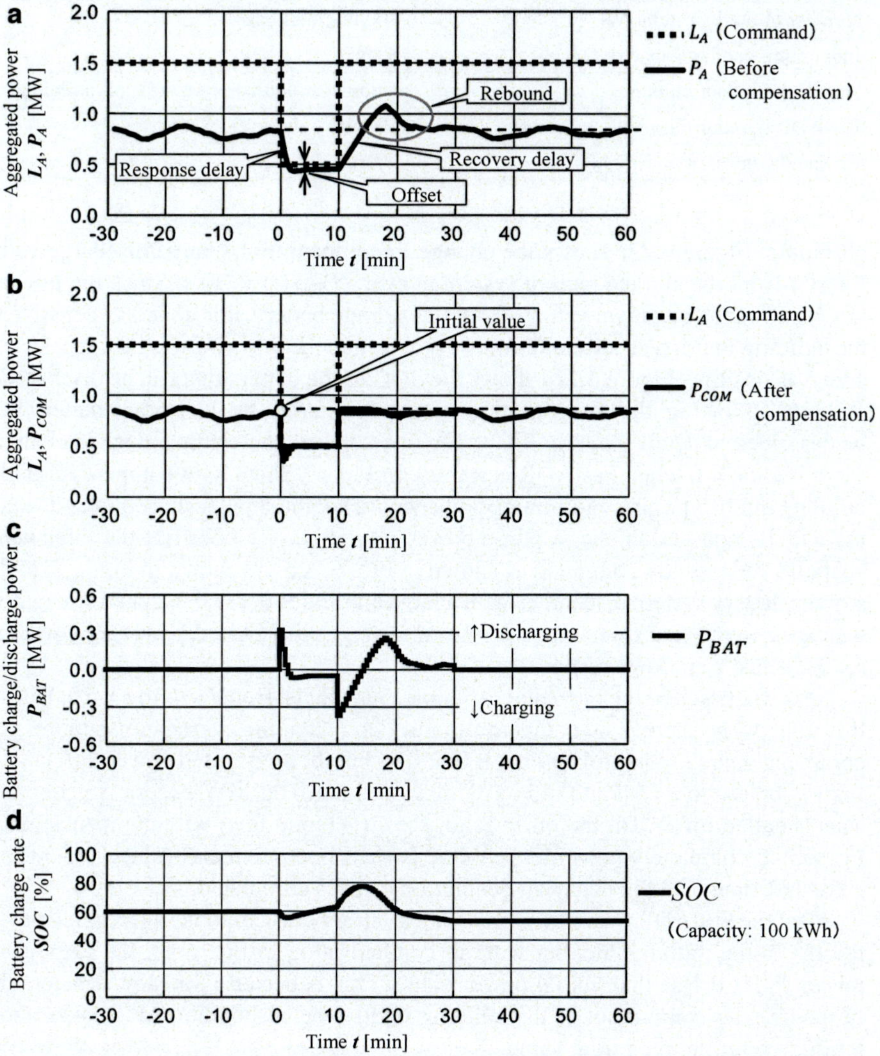

Fig. 6.22 Compensation for tracking the Negawatt power generation command by the storage battery of the building multi air-conditioning group FastADR

Table 6.2 Storage battery compensation simulation parameters for building multi air-conditioner group FastADR

Item	Value
Battery capacity C_{USA}	100 kWh
SOC range SOC_{MIN}~SOC_{MAX}	50–80%
Battery control cycle τ_{BC}	30 s
Battery control processing delay τ_{BP}	0 s
Outdoor unit rated cooling capacity Q_{ACMAX}	45 kW
Number of outdoor units N_C	100 unit/building
Indoor unit cooling capacity C_{Pi}	8 kW
Number of indoor units N_I	6 units per outdoor unit, 600 unit/building
FastADR duration T_{DR}	10 min
Rebound control period T_{REB}	20 min

obserable. Figure 6.22c shows the change in compensation charge/discharge value $P_{BAT}(t)$ from the storage battery system at every $P_{BAT}(t) = 30$ s. Since the upward graph shows the negative wattage from the storage battery, it is added as negative to the building multi-type air-conditioning power, and the downward is charged, so it is added as positive. Figure 6.22d shows the SOC of the battery that can be discharged.

In the following, the compensation status of FastADR by the storage battery will be described in detail. First, in Fig. 6.22b, the suppression command value changes stepwise at $t = 0$ when FastADR is started. In Fig. 6.22a without compensation, the building multi-type air-conditioning power could not the suppression conmand value, but with compensation, the response power value $P_{COM}(t)$ looks like the same wave shape as suppression command as shown in Fig. 6.22b. As shown in Fig. 6.22c, the storage battery system discharges with a large amplitude at $t = 0$ to generate negative wattage $P_{BAT}(t)$ and subtracting this from $P_A(t)$. Compensated power $P_{COM}(t)$ can realize a fast step-down response.

Next, we describe the FastADR duration T_{DR}, that is, from $t = 0$ to $t = 10$. During this time, when there is no compensation in Fig. 6.22a, the aggregation suppression command value $L_A(t)$ (broken line) is 0.5 MW, but the aggregated power $P_A(t)$ (solid line) is below to it. About 0.45 MW, that is, about 0.05 MW excess is generated. This is called offset. On the other hand, $P_{COM}(t)$ (solid line) with compensation in Fig. 6.22b completely coincides with the broken line command value $L_A(t)$ during FastADR time, and the negawatt generation error is eliminated.

Also, the FastADR recovery period, that is, after $t = 10$, until the aggregated power returns to the initial value, but without compensation in Fig. 6.20, the aggregated power $P_A(t)$ is less than initial power value. This is because the slow acceleration of the inverter compressor of the building multi-type air-conditioner. It takes about 5 min to return to the initial value.

Even though the resource aggregator commanded to stop the generation of negawatts, but it still generated negawatts in this period. On the other hand, in the case of compensation, as shown in Fig. 6.22a, it can follow the step release command completely at $t = 10$. This is because, as shown in Fig. 6.22b, the resource aggregator issued a control command to charge the storage battery and add power consumption to the response power.

Finally, let us focus on the rebound period immediately after the FastADR release. Here, the rebound recovery period T_{RP} after releasing the demand response is 20 min, which is twice the demand response duration T_{DR} (in this case, 10 min). In other words, in Fig. 6.22a and b, it was assumed that the response power $P_A(t)$ was required to be maintained at the FastADR start initial value (thin broken line) from $t = 10$ to $t = 30$.

As shown in Fig. 6.22a, when there is no compensation, the response power $P_A(t)$ starts FastADR for about 10 min from $t = 15$ to $t = 25$, regardless of the rebound recovery period after releasing FastADR. It has increased from the initial value of about 0.8 MW, that is, rebound has occurred. As shown in Fig. 6.22b, with compensation, the initial power value is accurately maintained during the rebound recovery period. This is because the resource aggregator monitors the aggregate response power $P_A(t)$ notified from the ERC gateway every control cycle $\tau_{BC} =$ 30 s, calculates the error from the aggregated power $P_A(t)$. In order to compensate, the charge/discharge command is issued to the battery system, $P_{BAT}(t)$ is generated and integrated with $P_A(t)$. So that the compensated power $P_{COM}(t)$ is maintained at the initial value.

In this way, by charging and discharging the storage battery system for FastADR, the initial SOC and the SOC after completion may differ depending on the case. In this example, during these series of operations, the amount of discharge was greater than the amount of charge, so the SOC decreased as a whole. When the rebound recovery period is over, it is necessary to gradually recover the SOC at a small rate to prepare for the next FastADR dispatch.

References

1. E. Mangalova, E. Agfonov, Time series forecasting using ensamble of AR models with time-varying structure, in *IEEE Conference on Evolving and Adaptive Intelligent Systems EAIS2012*, pp. 198–203 (2012)
2. Y. Iwata, K. Saito, C. Ninagawa, J. Morikawa, S. Kondo, Time series combined statistical modeling on dynamic response characteristics of ADR power curtailment for building facilities, in *IEEJ Industry Applications Society Conference*, pp. v265–v266 (2014) (in Japanese)
3. T. Naramura, J. Morikawa, C. Ninagawa, Dynamic characteristics of side effect of room temperature in FastADR control of multi-type building air-conditioners facilities. IEEJ Power Energy Soc. Conf. (2015) (in Japanese)
4. S. Kondo, J. Morikawa, C. Ninagawa, Averaging effect of characteristics on FastADR power curtailment response of multi-type building air-conditioning facilities. IEEJ Power Energy Soc. Conf. (2015) (in Japanese)
5. T. Fukazawa, Y. Iwata, J. Morikawa, C. Ninagawa, Stabilization of neural network by combination with AR model in FastADR control of building air-conditioner facilities. IEEJ Trans. Elec. Electron. Eng. **11**(1), 124–125 (2016)

6. T. Naramura, J. Morikawa, C. Ninagawa, Prediction model on room temperature side-effect due to FastADR aggregation for a cluster of building air-conditioning facilities. IEEJ Trans. Power Energy **136**(4), 432–438 (2016) (in Japanese)
7. S. Kondo, J. Morikawa, C. Ninagawa, Averaging effect on stochastic response prediction in FastADR aggregation for building air-conditioning facilities. IEEJ Trans. Electr. Electron. Eng. 11(6), 820–822 (2016)
8. Y. Nagata, T. Naramura, C. Ninagawa, J. Morikawa, Short time prediction model on room temperature side-effect due to fast demand response of building multi-type air-conditioning facilities power consumption. IEEJ Nat. Convention **4**, 390–391 (2016). ((in Japanese))
9. A. Nakamura, S. Kondo, C. Ninagawa, J. Morikawa, Time-series data mining for characteristics model on fast demand response of building multi-type air-conditioning facilities power consumption. IEEJ Nat. Convention **4**, 392–393 (2016). ((in Japanese))
10. K. Suzuki, C. Ninagawa, J. Morikawa, T. Inaba, S. Kondo, Estimation on averaging effect of large-scale aggregation of DR transient responses of building multi-type air-conditioning facilities. IEEJ Trans. Power Energy **138**(7), 582–590 (2018) (in Japanese)
11. A. Kiyota, M. Takahama, C. Ninagawa, Wide area network discrete feedback control on fast ADR of a cluster of building air-conditioning facilities. IEEJ Trans. Electr. Electron. Eng. **11**(6), 826–828 (2015)
12. H. Asaka, M. Takahama, C. Ninagawa, Mitigation of saturated LFC thermal power plants by very FastADR aggregation of a huge number of building multi-type air-conditioners. IEEJ Trans. Electr. Electron. Eng. **12**(3), pp.442–443, 2017
13. C. Ninagawa, H. Asaka, M. Takahama, Asymmetric constant-rate transfer function model for FastADR feedback control of cluster of building multi-type air-conditioners. IEEJ Trans. Power Energy **137**(8), 566–572, 2017 (in Japanese)
14. J. Morikawa, S. Kondo, S. Isozumi, C. Ninagawa, New fast demand control of building HVAC facilities for smart grid. Mitsubishi Heavy Indus. Tech. Rev. **51**(2), 10–15 (2014) (in Japanese)
15. C. Ninagawa, Automated demand response aggregation and building energy management system. JSRE Annual Conf. D331 (2015) (in Japanese)

Chapter 7
Virtual Power Plant Performance

7.1 Building Multi-type Air-Conditioning Power Control Model

7.1.1 Power Consumption of Building Multi-type Air-Conditioning

The refrigerant compressor accounts for 90% of the power consumption of building multi-type air-conditioning. Therefore, in this chapter, the power consumption of the refrigerant compressor is modeled based on the principle of the refrigeration cycle.

Figure 7.1 shows the refrigerant circuit for building multi-type air-conditioning. The refrigerant circuit is filled with refrigerants such as R410A and R32. In the cooling operation, while the refrigerant circulates in the refrigerant circuit, the indoor heat exchanger absorbs the indoor heat and the outdoor heat exchanger discharges the heat to the outside. The refrigeration cycle realizes this mechanism.

At this time, an appropriate compressor speed is determined as needed while constantly monitoring the refrigerant pressure and temperature so that the microcomputer can properly maintain the refrigeration cycle.

Figure 7.2 is the Mollier diagram, which shows how the refrigerant temperature, pressure, phase (liquid phase/wet vapor/gas phase), and thermal energy change during the refrigeration cycle. The horizontal axis of this Mollier diagram is the specific enthalpy h [kJ/kg] and the vertical axis is the pressure p [kPa]. Specific enthalpy is the thermal energy contained in the refrigerant per unit mass and is expressed by the Eq. 7.1.

$$h = u + pv \tag{7.1}$$

where u is the specific internal energy [kJ], p is the pressure [kPa], and v is the specific volume [m^3/kg].

C. Ninagawa, *Virtual Power Plant System Integration Technology*, Power Systems,
https://doi.org/10.1007/978-981-16-6148-8_7

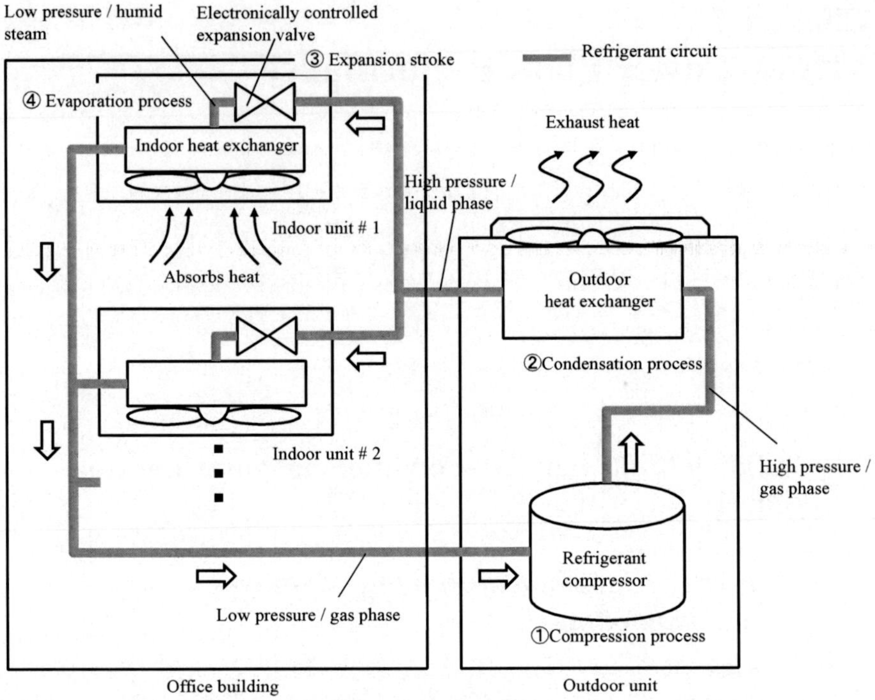

Fig. 7.1 Building multi air-conditioning refrigerant circuit

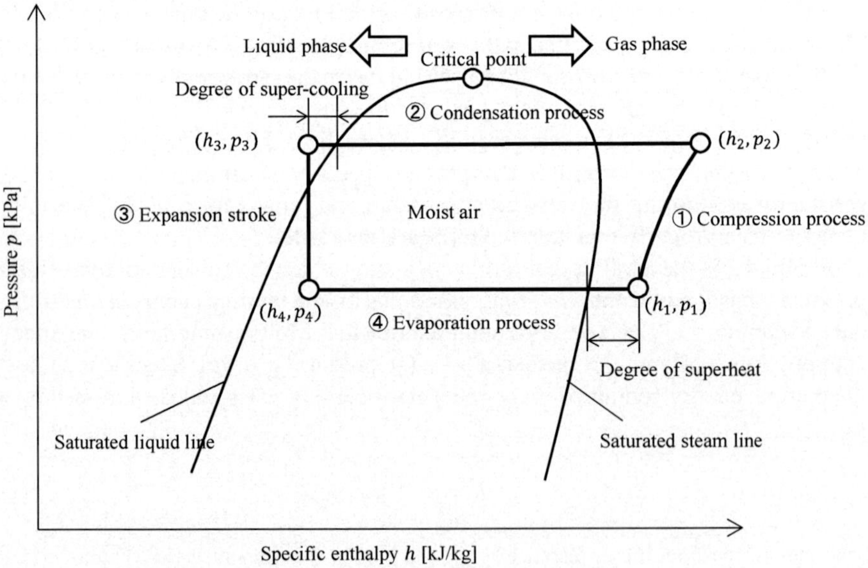

Fig. 7.2 Mollier diagram and refrigeration cycle

The refrigerant follows the following path on the Mollier diagram (①–④ in Figs. 7.1 and 7.2).

① Compression process
On the suction side of the compressor, the refrigerant is a low-temperature, low-pressure gas. The refrigerant is compressed by the compressor into a high-temperature, high-pressure gas.

② Condensation process
The high-temperature and high-pressure refrigerant that has undergone the compression process undergoes a phase change by discharging sensible heat and latent heat to the outside through the outdoor heat exchanger, and becomes a high-temperature and high-pressure liquid. At this time, there is a slight sensible heat change even after it becomes liquid.

③ Expansion stroke
When the refrigerant passes through the electronically controlled expansion valve of the indoor unit, the high-temperature and high-pressure liquid changes from low-temperature and low-pressure humid air (a mixture of liquid and gas). The microcomputer of the indoor unit adjusts the opening of the electronically controlled expansion valve while monitoring the refrigerant temperature so that the refrigerant can evaporate in the evaporation process described later.

④ Evaporation process
This is the actual cooling process. When the refrigerant passes through the indoor heat exchanger, it takes away the latent heat of vaporization and turns it into a low-temperature, low-pressure gas that lowers the ambient air. Since the refrigerant is still low in temperature after becoming a gas, it is accompanied by a slight sensible heat change. This is called superheat.

As shown in Fig. 7.2, the refrigerant compressor not only circulates the refrigerant but also compresses the refrigerant from a low temperature/low pressure gas to a high temperature/high pressure gas.

The amount of work performed by this refrigerant compressor per unit refrigerant mass is equal to the change in the compression stroke enthalpy $\Delta h_{12} = h_2 - h_1$ [kJ/kg] on the Mollier diagram. Therefore, the required power W [kW] for the compression stroke is given by Eq. 7.2 if the refrigerant circulation rate is G [kg/s].

$$W = \Delta h_{12} G \tag{7.2}$$

Further, the compression stroke enthalpy change Δh_{12} [kJ/kg] is obtained by Eq. 7.3, assuming that the compression stroke is adiabatic compression with no external entry and exit of heat and the refrigerant is an ideal gas.

$$pv^{\kappa} = p_1 v_1^{\kappa} = p_2 v_2^{\kappa} = const \tag{7.3}$$

The above is Poisson's law. Where p_1 is the compressor suction pressure [kPa], p_2 is the compressor discharge pressure [kPa], v_1 is the suction specific volume [m^3], and v_2 is the discharge specific volume [m^3]. κ is the refrigerant specific heat ratio, which is the ratio C_p/C_v between the constant pressure specific heat C_p [kJkg^{-1} K^{-1}] and the constant volume specific heat C_v [kJkg^{-1} K^{-1}].

Now, when work is applied to the sealed gas and the pressure and volume change from p_1, v_1 to p_2, v_2, the work size becomes the area on the p–v diagram shown in Fig. 7.3. Using Poisson's law, the Eq. 7.4 is obtained.

$$\Delta h_{12} = \int_{v_2}^{v_1} p dv = p_1 v_1^{\kappa} \int_{v_2}^{v_1} \frac{1}{v^{\kappa}} dv = \frac{p_1 v_1^{\kappa}}{\kappa - 1}\left(\left(\frac{v_1}{v_2}\right)^{\kappa-1} - 1\right) \tag{7.4}$$

Furthermore, the equation of state of an ideal gas is expressed as Eq. 7.5 using the gas constant R [JK^{-1} mol^{-1}] and temperature T [K].

$$pv = RT \tag{7.5}$$

Therefore, the Eq. 7.6 is established from Eqs. 7.3 and 7.5.

$$T v^{\kappa-1} = T_1 v_1^{\kappa-1} = T_2 v_2^{\kappa-1} = const \tag{7.6}$$

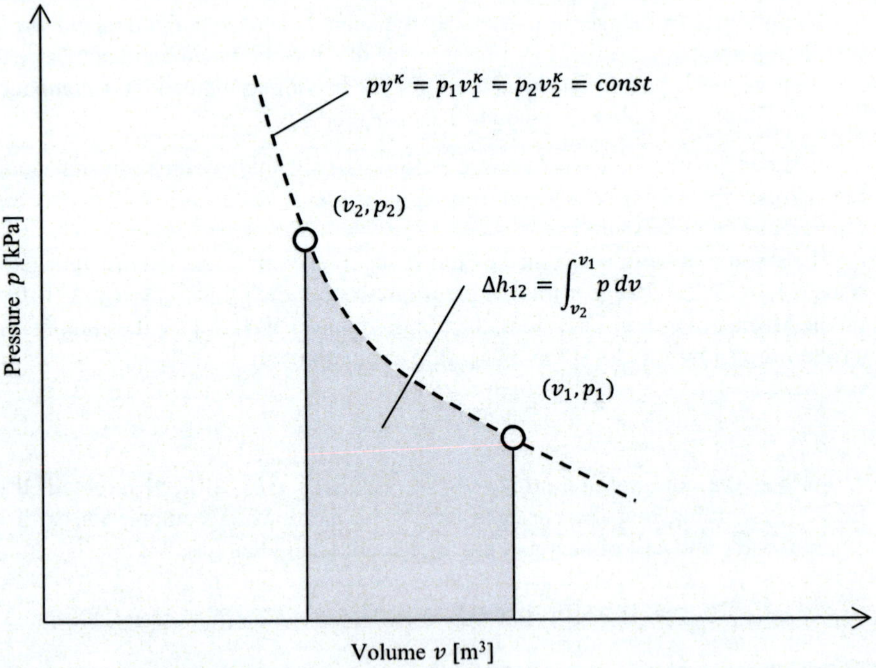

Fig. 7.3 *P–V* diagram

$$\frac{T}{p^{\frac{\kappa-1}{\kappa}}} = \frac{T_1}{p_1^{\frac{\kappa-1}{\kappa}}} = \frac{T_2}{p_2^{\frac{\kappa-1}{\kappa}}} = const \tag{7.7}$$

From Eqs. 7.6 and 7.7, 7.4 become as Eq. 7.8.

$$\Delta h_{12} = \frac{p_1 v_1^{\kappa}}{\kappa - 1}\left(\left(\frac{p_2}{p_1}\right)^{\frac{\kappa-1}{\kappa}} - 1\right) \tag{7.8}$$

The above is the enthalpy difference corresponding to the required power of the compressor per unit refrigerant circulation rate.

Actually, heat enters and exits during the compression process, so the Eq. 7.9 similar to Poisson's law is used.

$$pv^n = p_1 v_1^n = p_2 v_2^n = const \tag{7.9}$$

The compression process according to Eq. 7.9 is called polytropic compression, and n is called the polytropic index. Note that $n = 1$ is isothermal, $n = \kappa$ is adiabatic, and polytropic compression is $1 < n < \kappa$. In other words, when the compression process is polytropic compression, κ is calculated as n in Eq. 7.4.

Subsequently, the refrigerant circulation rate G [kg/s] is obtained by Eq. 7.10.

$$G = \frac{\eta_v v_{cmp} N}{v_1} \tag{7.10}$$

η_v is the compressor volumetric efficiency (0.0–1.0), v_{cmp} is the compressor displacement [m^3/rev], and N is the compressor speed [rps]. Furthermore, the compressor power consumption P [kW] is the required power physics model, assuming that the motor efficiency is η_m (0.0–1.0), so that the power consumption of the refrigerant compressor, which accounts for the majority of the power consumption of the building multi-type air-conditioner, is Derived based on Eq. 7.11.

$$P = \frac{W}{\eta_M} \tag{7.11}$$

Next, the air-conditioning capacity Q_{AC} [kW] that the indoor unit heat exchanger exhibits in total is derived. Q_{AC} [kW] can be obtained as follows by multiplying the evaporation process enthalpy change $\Delta h_{41} = h_1 - h_4$ by the refrigerant circulation rate G [kg/s].

$$Q_{AC} = \Delta h_{41} G \tag{7.12}$$

The Eq. 7.13 is called Coefficient Of Performance (COP) η_{cop}, which represents the energy efficiency of air-conditioning.

$$\eta_{cop} = \frac{Q_{AC}}{P} \tag{7.13}$$

The air-conditioning capacity Q_{ACi} [kW] of each indoor unit side heat exchanger is calculated by proportionally allocating the total air-conditioning capacity Q_{AC} [kW] of all indoor units according to the rated capacity C_{Pi} [kW] of the indoor unit.

$$Q_{ACi} = \frac{S_{THi}C_{Pi}}{\sum_{i=1}^{N_I}(S_{THi}C_{Pi})}Q_{AC} \tag{7.14}$$

S_{THi} is the thermo on/off state (0 or 1), and N_I is the number of indoor units. Thermo on/off is a control that switches on/off the air-conditioning of the indoor unit according to the difference between the set room temperature and the measured room temperature, which will be explained in the next section.

In the building multi-type air-conditioning model described in this chapter, the total capacity of the thermo-on indoor units is set as the target air-conditioning capacity Q^*_{AC} [kW], and the refrigerant circulation amount G^* [kg/s] that satisfies the air-conditioning capacity flows through the refrigerant circuit. The compressor accelerates/decelerates. At this time, the target power consumption P^* [kW] of the compressor is obtained as Eqs. (7.15–7.17).

$$Q^*_{AC} = \sum_{i=1}^{N_I}(S_{THi}C_{Pi}) \tag{7.15}$$

$$G^* = \frac{Q^*_{AC}}{\Delta h_{41}} \tag{7.16}$$

$$P^* = \frac{\Delta h_{12}G^*}{\eta_M} \tag{7.17}$$

7.1.2 Building Multi-type Air-Conditioning Power Control Model

In this section, the air-conditioning power and room temperature dynamic physical models [1] and [2] for the building multi-type air-conditioning power limit command are modeled by the difference equation as Eq. 7.18.

$$P(t + \Delta t) = P(t) + \Delta P(t)\Delta t \tag{7.18}$$

$$\Delta P(t) = S^{PL}_{OV}D_{DWN} + \left(1 - S^{PL}_{OV}\right)\left\{S^{TH}_{OV}D_{DWN} + \left(1 - S^{TH}_{OV}\right)S^{PL}_{FR}S^{TH}_{FR}D_{UP}\right\} \tag{7.19}$$

$$T_{Ai}(t + \Delta t) = T_{Ai}(t) + \Delta T_{Ai}(t)\Delta t \tag{7.20}$$

$$\Delta T_{Ai}(t) = \frac{1}{C_{Hi}}(Q_{Li}(t) - Q_{ACi}(t)) \tag{7.21}$$

where t is the discrete time [s] and the model operates with a sampling width Δt [s]. $P(t)$ is power consumption [kW], $\Delta P(t)$ is the rate of change in power consumption [kW/s], $T_{Ai}(t)$ is the measured room temperature [°C] of indoor unit i, and $\Delta T_{Ai}(t)$ is the change in room temperature rate [°C/s].

First, the difference equation for power consumption is described. This model changes the power consumption $P(t)$ according to the power suppression command value $L(t)$ and the temperature control required power $P^*(t)$ [kW]. The required temperature control power $P^*(t)$ is the target value of power consumed by the air-conditioner when the power suppression command value $L(t)$ is released ($L(t) = \infty$). Is described in the next section.

In Eq. 7.19, D_{UP} is the power increase rate [kW/s] and D_{DWN} is the power decrease rate [kW/s]. S_{OV}^{PL} and S_{FR}^{PL} are control statuses indicating the relationship between the power suppression command value $L(t)$[kW] and power consumption $P(t)$, and take two values (0 or 1) according to the conditions shown in Table 7.1. k_L in the table is the power consumption tracking coefficient with respect to the power suppression command value. S_{OV}^{TH} and S_{FR}^{TH} are control statuses indicating the relationship between the temperature control required power $P^*(t)$ and the power consumption $P(t)$, and change as shown in Table 7.2.

Figure 7.4 shows the relationship between power consumption and control status. In Fig. 7.4, the model changes as ①–④. First, ① If the current power consumption exceeds the power suppression command value $L(t)$, it will drop to the power suppression command value $L(t)$. When the power suppression command value $L(t)$ is increased, the power consumption $P(t)$ increases following the power suppression command value $L(t)$. (3) When the power suppression command value $L(t)$ exceeds the required temperature control power $P^*(t)$, the power consumption $P(t)$ increases to the required temperature control power $P^*(t)$. (4) When the required temperature

Table 7.1 Control status related to power suppression command value $L(t)$

Condition control status	$P(t) > k_L L(t)$	$P(t) = k_L L(t)$	$P(t) < k_L L(t)$
S_{OV}^{PL}	1	0	0
S_{FR}^{PL}	0	0	1

Table 7.2 Control status related to temperature control power $P^*(t)$

Condition control status	$P(t) > P^*(t)$	$P(t) = P^*(t)$	$P(t) < P^*(t)$
S_{OV}^{TH}	1	0	0
S_{FR}^{TH}	0	0	1

S_{OV}^{PL}	1	0	0	0	0	0	0	0
S_{FR}^{PL}	0	0	1	0	1	1	1	1
S_{OV}^{TH}	0	0	0	0	0	0	1	0
S_{FR}^{TH}	1	1	1	1	1	0	0	0
ΔP	D_{DWN}	0	D_{UP}	0	D_{UP}	0	D_{DWN}	0

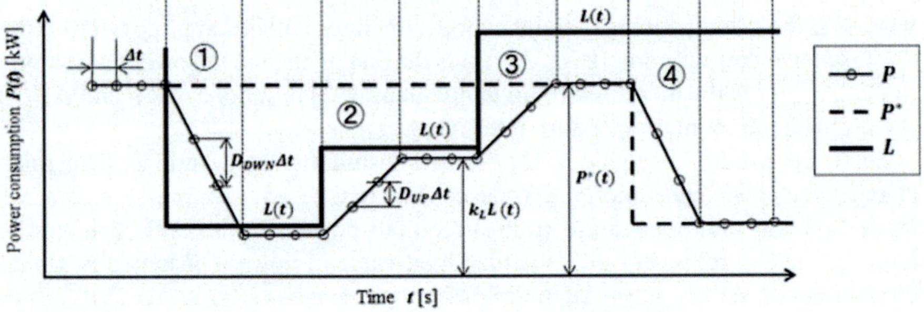

Fig. 7.4 Relationship between power consumption and control status

control power $P^*(t)$ decreases, the power consumption $P(t)$ decreases to the required temperature control power $P^*(t)$. $P(t)$ transitions with the rated power consumption P_{MAX} [kW] as the upper limit and the minimum power consumption P_{MIN} [kW] as the lower limit.

The difference equation at room temperature is described as Eq. 7.20. The difference equation for room temperature $T_{Ai}(t)$ exists independently for each indoor unit i. In other words, different room temperature dynamics are simulated for each indoor unit. In Eq. 7.21, C_{Hi} is the heat capacity [kJ/°C], $Q_{Li}(t)$ is the heat load [kW], $Q_{ACi}(t)$ is the air-conditioning capacity [kW] of the indoor unit.

The air-conditioning capacity $Q_{ACi}(t)$ of the indoor unit is distributed to the indoor unit by Eq. 7.22 using the air-conditioning capacity $Q_{AC}(t)$ [kW] exhibited by the outdoor unit.

$$Q_{AC}(t) = \eta_{COP} P(t) \tag{7.22}$$

$$Q_{ACi} = \frac{S_{THi} C_{Pi}}{\sum_{i=1}^{N_I} (S_{THi} C_{Pi})} Q_{AC} \tag{7.23}$$

Here, η_{COP} is called the coefficient of performance (COP) and represents the air-conditioning capacity [kW] per 1 [kW] of power consumption. S_{THi} is the thermo on/off status (0 or 1), C_{Pi} i is the rated capacity [kW] of the indoor unit, and N_I is the number of indoor units.

Thermo on/off is a control to switch on/off the air-conditioning of the indoor unit based on the difference between the set room temperature T_{Si} and the measured room temperature $T_{Ai}(t)$ As shown in Fig. 7.5, when the measured room temperature T_{Ai} (t) of the indoor unit reaches $T_{Si} - \Delta T_{TH\_OFF}$, the thermo unit is turned off ($S_{THi} = 0$),

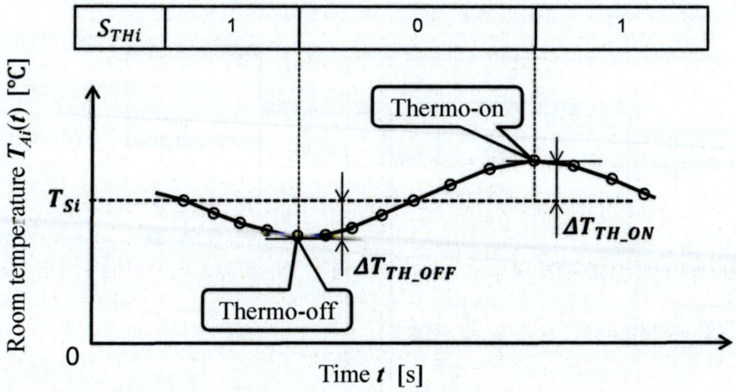

Fig. 7.5 Thermo ON/OFF state

Fig. 7.6 Thermo-ON/OFF hysteresis

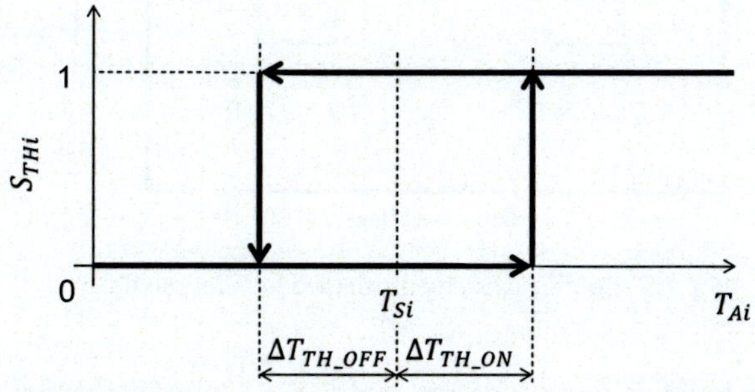

and the air-conditioning of the indoor unit stops. When the room temperature rises and reaches $T_{SETi} + \Delta T_{TH\_OFF}$, thermo-on ($S_{THi} = 1$) and restart air-conditioning. Here, ΔT_{TH_ON} and ΔT_{TH_OFF} are the hysteresis width [°C], and the state of S_{THi} is shifted only in the direction of the arrow as shown in Fig. 7.6 to prevent frequent thermo-on/off.

The cooling capacity distributed to indoor units will be explained using Fig. 7.7 as an example. In Fig. 7.7, six indoor units are connected to the outdoor unit. Now, when 3 out of 6 indoor units are thermo-on, the cooling capacity exerted by the outdoor unit is distributed to the 3 indoor units. Refrigerant does not flow into the remaining three units that are thermo-off, and cooling capacity is not distributed.

The cooling capacity of the outdoor unit is distributed to the thermo-on indoor unit by the ratio of the rated capacity of the indoor unit. In the case of Fig. 7.7, the cooling capacity of the outdoor unit is $Q_{AC}(t) = 14$ [kW], and the ratio of the indoor unit rated capacity is $C_{P1} : C_{P2} : C_{P3} = 4[\text{kW}] : 8[\text{kW}] : 16[\text{kW}] = 1 : 2 : 4$. Therefore, the indoor unit cooling capacity is distributed to $Q_{AC1} : Q_{AC2} : Q_{AC3} = 1 : 2 : 4 = 2[\text{kW}] : 4[\text{kW}] : 8[\text{kW}]$ and modeled.

In this way, the change in air-conditioning power $P(t)$ and the accompanying change in capacity to each air-conditioner, the change in each room temperature $T_{Ai}(t)$, and each setting in response to the power restriction command $L(t)$ of the

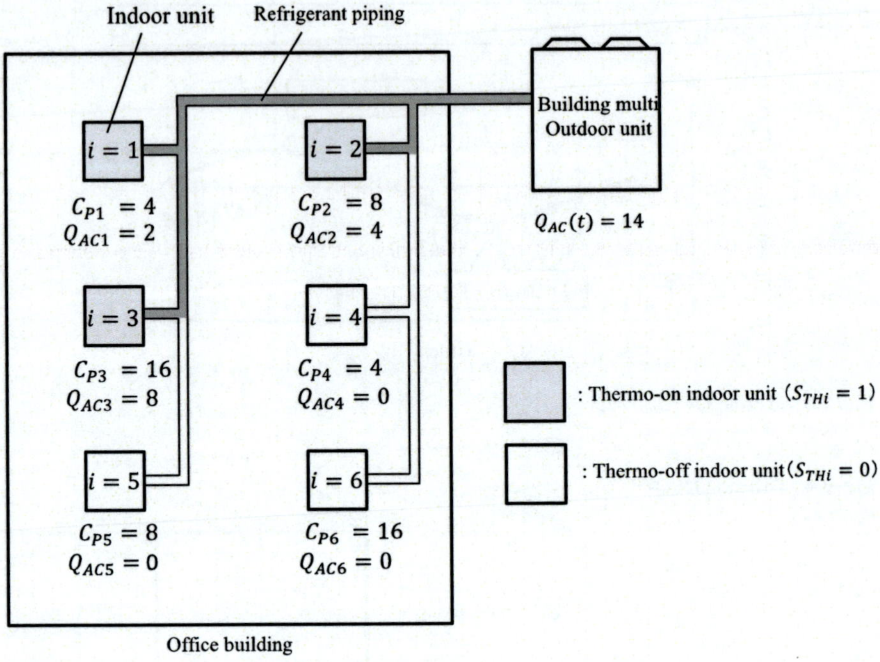

Fig. 7.7 Cooling capacity distributed to indoor units

building multi-type air-conditioner A physical model of thermo-on/off behavior with temperature is obtained.

7.2 Office Building Thermal Load Simulation Model

7.2.1 Solar Heat Load Model

In the previous section, the physical model of the power limiting performance on the building multi-type air-conditioner side was shown, but in this section, the physical model of the heat load on the air-conditioning versus the building side is shown. The walls and ceiling of the office building are heated by solar radiation and become an indoor heat load. The temperature of the outer wall heating by solar radiation as if the outside air temperature has risen is called the equivalent outside air temperature *SAT*. First, the heat q [kW/m^2] transferred per unit area on the outer surface is as Eq. 7.24.

$$q = aI_W + \alpha_O (T_{OUT} - T_{WALL}) \tag{7.24}$$

where a is the solar absorptivity (0.0–1.0), I_W is the solar radiation [kW/m^2] irradiated to the wall, α_O is the outer surface heat transfer coefficient [kW/(m^2 °C)], and T_{WALL} is the outer wall It is the temperature [°C]. When this equation is expressed as α_O, it becomes as Eq. 7.25.

$$q = \alpha_O\left\{\left(\frac{a}{\alpha_O} I_W + T_{OUT}\right) - T_{WALL}\right\} \tag{7.25}$$

In Eq. 7.25, terms related to solar radiation and outside air temperature are collectively referred to as *SAT*.

$$SAT = \frac{a}{\alpha_O} I_W + T_{OUT} \tag{7.26}$$

At this time, if the heat transfer delay due to the heat capacity of the outer wall is not taken into account, the following relationship holds between the heat received by the outer surface and the heat transferred to the room.

$$q = \alpha_O(SAT - T_{WALL}) = K_W(SAT - T_{AIR}(t)) \tag{7.27}$$

Here, K_W is the heat transfer rate of the wall [kW/(m^2 °C)].

The solar radiation I_W reaching the outer wall is calculated based on the global solar radiation I_h [kW/m^2], which is the solar radiation on the ground plane. As shown in Fig. 7.8, total solar radiation I_h is direct solar radiation I_s [kW/m^2], which is highly directional from the sun, and sky solar radiation I_v [kW/m^2].

$$I_h = I_s \sin\theta_{sun} + I_v \tag{7.28}$$

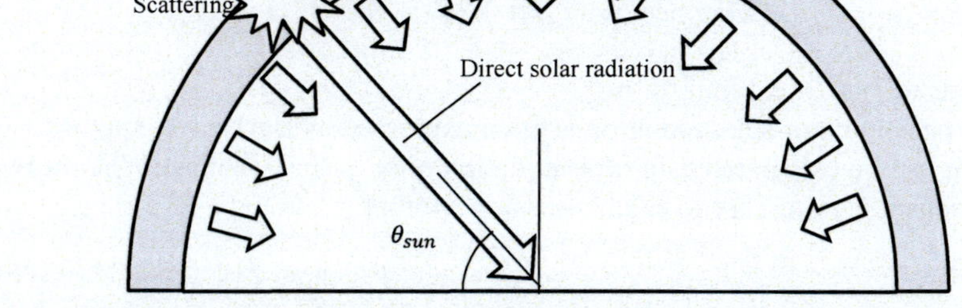

Fig. 7.8 Direct solar radiation and sky solar radiation

Here, θ_{sun} [deg] is called the solar altitude and is an angle determined by the latitude and time of the building. The approximate direct solar radiation I_s and sky solar radiation I_v at a certain time can use the solar radiation data of neighboring points published on the Japan Meteorological Agency website [3]. The solar radiation data provided by the Japan Meteorological Agency is only global solar radiation I_h except for specific observation points, and the ratio of direct solar radiation I_s and sky solar radiation I_v is unknown. Therefore, calculation is performed by separating direct solar radiation I_s and sky solar radiation I_v by a technique called direct scattering separation.

For the expression of direct solar radiation, the following theoretical formula, the Bougues formula, is well known. Bouguer's equation summarizes the attenuation rate that varies depending on the wavelength of sunlight in the atmosphere as the atmospheric transmittance. Also, it does not include the effects of shadows and weather on the surface.

$$I_s = I_O^{'}\varepsilon_{AIR}^{m} \tag{7.29}$$

$$I_O^{'} = \frac{I_O}{r^2} \tag{7.30}$$

$$\begin{aligned} m &= \frac{1}{\sin\theta_{sum}} \text{ (Ignore curvature)} \\ &= \sqrt{(797\sin\theta_{sum})^2 + 1595} - 797\sin\theta_{sum} \end{aligned} \tag{7.31}$$

where I_O is the solar constant (1.37 [kW/m^2]), ε_{AIR} is the atmospheric transmittance (0.0–1.0), and r is the solar radius [−]. m is the air mass (atmospheric mass), and is the relative value when the atmospheric mass involved in the scattering and absorption of solar radiation is assumed to be 1 when solar radiation penetrates the atmospheric layer vertically [3]. In general, the atmospheric mass m is calculated using an expression that ignores the curvature of the earth.

Although ε_{AIR} depends on the degree of air pollution and humidity, the approximate amount of total solar radiation at a representative point in Japan (Kiyose City, Tokyo) can be obtained using Eqs. 7.32 and 7.33.

$$\begin{cases} \text{July:} \quad \varepsilon_{AIR} = 0.670 + 0.0040(\tau - 12)^2 \\ \text{October:} \quad \varepsilon_{AIR} = 0.740 + 0.0045(\tau - 12)^2 \end{cases} \tag{7.32}$$

where τ is the 24 h time [h].

The amount of solar radiation is often expressed by Berlagage equation Eq. 7.34, but it can be calculated using this equation only in fine weather. When it is cloudy or rainy, it is necessary to calculate using another relational expression.

$$I_v = 0.5\sin\theta_{sum} I_O^{'} \frac{1 - \varepsilon_{AIR}^{m}}{1 - 1.4\ln\varepsilon_{AIR}} \tag{7.34}$$

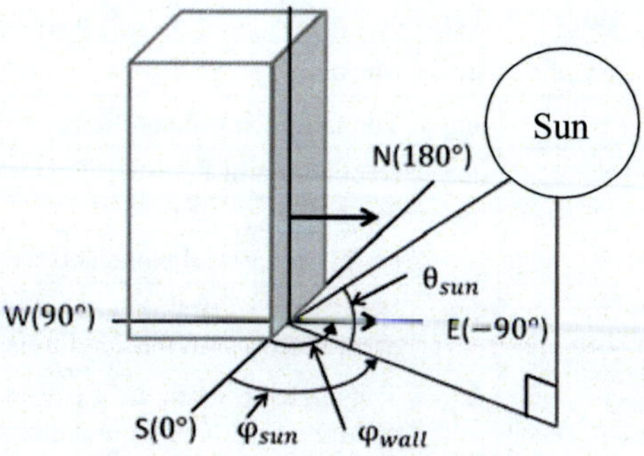

Fig. 7.9 Solar altitude and direction

For direct dispersion separation, the observed value is assigned to the left side of Eq. 7.32, and the right side is calculated using Eq. 7.33. Direct solar radiation I_s and sky solar radiation I_v can be calculated by substituting the solutions obtained in this way into each equation. The solar radiation data published by the Japan Meteorological Agency is the cumulative total solar radiation amount J_h [1 h before] Since [MJ/m^2], the unit is converted to [kW/m^2] by the following Eq. 7.35.

$$I_h = \frac{J_h \cdot 10^3}{3600} \tag{7.35}$$

Since the total solar radiation I_h is the amount of solar radiation that irradiates the ground plane, as shown in Fig. 7.9, the solar radiation $I_w\left[\mathrm{kW/m^2}\right]$ that irradiates the vertical plane of the actual building is It depends on the direction and is expressed as Eq. 7.36.

$$I_w = I_s \cos\theta_{sun} \cos(\varphi_{sun} - \varphi_{wall}) + \frac{I_v}{2} \tag{7.36}$$

where ϕ_{sum} is the solar azimuth [deg] and φ_{wall} is the building wall azimuth [deg]. The azimuth is set to 0 for the south direction, negative [−] for the eastbound direction, and positive [+] for the westbound direction. The solar altitude θ_{sun} and the solar azimuth angle ϕ_{sum} have the relationship shown in Fig. 7.9 and can be obtained from the following equations using the parameters shown in Table 7.3.

$$\theta_{sun} = \arcsin(\sin\gamma \sin\delta + \cos\gamma \cos\delta \cos t)$$

$$\varphi_{sun} = \arccos\left(\frac{\sin\theta_{sun} \sin\gamma - \sin\delta}{\cos\theta_{sun} \cdot \cos\gamma}\right)$$

$$\delta = \arcsin(\cos(\tau + \mu)\sin(\delta_w))$$

$$t = 15(T - 12) + (\varepsilon - 135) + E_t$$

Table 7.3 Parameter list for solar radiation calculation

Symbol	Unit	Meaning
γ	deg	The latitude of the building
δ	deg	Declination of the sun (apparent height from the sun's equator)
t	deg	Time angle (angular distance from east to west along the intersection of the earth's equatorial plane and the celestial sphere with respect to the meridian)
τ	deg	Near-point separation angle (an angle measured from the position at which the earth is closest to the sun to the current position in the revolution direction)
μ	deg	The angle between the position where the earth is closest to the sun on its orbit and the position where the northern hemisphere of the earth is in the winter solstice
δ_w	deg	Day declination (≒ 23.44) at the winter solstice in the Northern Hemisphere
T	hour	Standard time for calculation
ε	deg	The longitude of the building
E_t	min	Time difference (difference in movement between the average sun, which is supposed to move at a constant speed on the celestial sphere, and the visible sun)
M	deg	The angle from the day when the earth is closest to the current position when it is assumed that the earth moves on the revolution orbit at the same angular velocity
T_{orbit}	day	Number of revolution cycles based on the day when the earth is closest to the sun (365.2596)
D	day	Year-round day of the calculation date with January 1 as D = 1
Y	year	The year number on the target date

$$\tau = M + 1.914 \sin M + 0.02 \sin 2M$$

$$\mu = 12.3901 + 0.0172\left(\mathrm{n} + \frac{M}{360}\right)$$

$$E_t = E_{t1} - E_{t2}$$

$$E_{t1} = M - \tau$$

$$E_{t2} = \arctan\left(\frac{0.043 \sin 2(\tau + \mu)}{1 - 0.043 \cos 2(\tau + \mu)}\right)$$

$$M = 360\frac{D - d_0}{T_{orbit}} C_{Hi}\frac{dT_{AIRi}(t)}{dt} = Q_{Li}(t) - Q_{ACi}(t)$$

$$d_0 = 3.71 + 0.2596n - INT\left[\frac{n+3}{4}\right] \quad n = Y - 1968$$

7.2.2 Thermal Load Model of Standard Office Building

As shown in Eq. 7.21, the room heat capacity C_{Hi}[kJ/°C], the room heat load $Q_{Li}(t)$ [kW], the air-conditioning capacity $Q_{ACi}(t)$ [kW], and the room temperature $T_{AIRi}(t)$ [°C], The following heat balance relationship holds.

$$C_{Hi}\frac{T_{Ai}(t)}{dt} = Q_{Li}(t) - Q_{ACi}(t) \tag{7.37}$$

That is, room temperature responds slowly in rooms with relatively large heat capacity C_{Hi}, and responds quickly in small rooms. In a room with a large thermal load $Q_{Li}(t)$, the room temperature rises quickly when the thermostat is off ($Q_{ACi}(t) = 0$), and in a room with a small $Q_{Li}(t)$, the room temperature rises slowly. Thus, the dynamic characteristics at room temperature change depending on the heat capacity C_{Hi} and the heat load $Q_{Li}(t)$, so these must be estimated reasonably.

In this section, C_{Hi} and $Q_{Li}(t)$ are clarified by constructing a building heat balance model for the section in charge of one indoor unit of the hypothetical office building shown in Fig. 7.10. $Q_{Li}(t)$ is composed of the Eq. 7.38.

$$Q_{Li}(t) =_O (t) + Q_R(t) + Q_I(t)A_R = L_C \cdot S_C \tag{7.38}$$

where $Q_O(t)$ is the intrusion heat load from the outer wall, floor, and ceiling [kW], $Q_R(t)$ is the solar radiation load that passes through the window glass [kW], and $Q_I(t)$ is the internal heat load [kW]. It is. Adjacent compartments are assumed to be air-conditioned, and there is no Q_N of heat in and out of the compartments.

(1) **Calculation of heat capacity of indoor unit air-conditioning section**

First, obtain the heat capacity C_{Hi} of the air-conditioning section of each indoor unit. When the air-conditioner is in operation, it cools not only room air but also walls and ceilings at the same time. Therefore, in this book, the total heat capacity of room air,

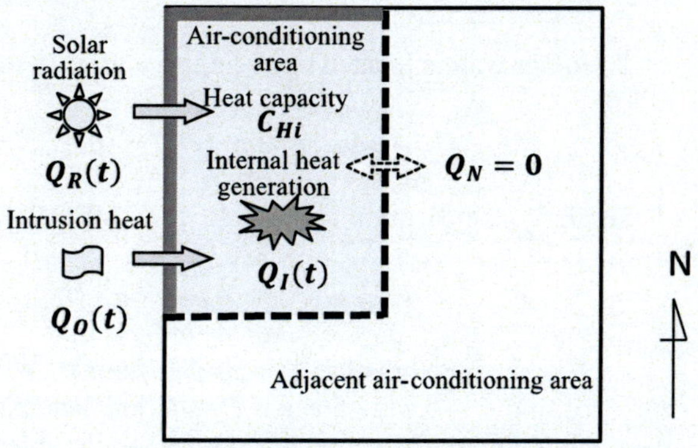

Fig. 7.10 Fictitious office for heat balance calculation

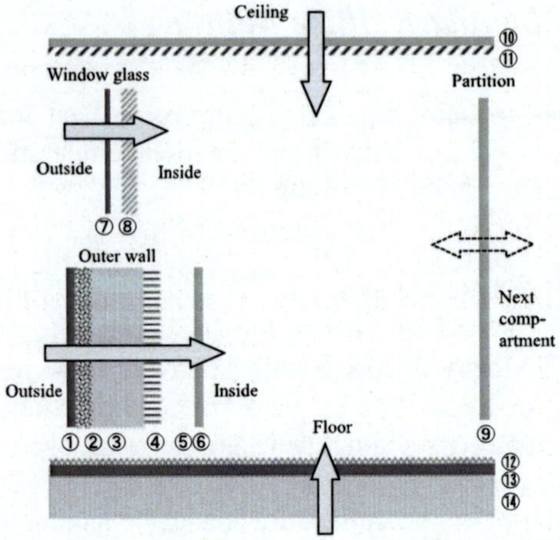

Place	Name of substance
Outer wall	① Tile
	② Cement
	③ Ordinary concrete
	④ Styro-foam
	⑤ Non-hermetic hollow layer
	⑥ Plasterboard
Window glass	⑦ Clear glass
	⑧ Light blinds
Partition	⑨ Plasterboard
Ceiling	⑩ Plasterboard
	⑪ Rock wool sound absorbing board
Floor	⑫ Carpet
	⑬ Floor P tile
	⑭ Ordinary concrete

Fig. 7.11 Example of building materials for a fictitious office building

floor, outer wall, window glass, ceiling, and partition is used as the heat capacity of the compartment.

In addition, this fictitious office building is composed of the members shown in Fig. 7.11 as an example. The room dimensions are vertical L_C[m], horizontal S_C [m], and the height to the ceiling H_C [m]. It was assumed that the west side of the room consisted of an outer wall and window glass, only the outer wall on the north side, and a partition on the south side.

At this time, the floor surface area A_R [m^2] is as in Eq. 7.39.

$$A_R = L_C \cdot S_C \tag{7.39}$$

Next, the window glass area A_G [m^2] is as in Eq. 7.40, assuming that the height of the window glass is H_G [m].

$$A_G = L_C \cdot H_G \tag{7.40}$$

The outer wall is located on the north and west sides, and its area A_W [m^2] is

$$A_W = L_C \cdot (H_C - H_G) + S_C \cdot H_C. \tag{7.41}$$

Similarly, the partition area A_P [m^2] is as Eq. 7.42.

$$A_P = S_C \cdot H_C \tag{7.42}$$

Each heat capacity [kJ/°C] of air C_A, floor C_{NF}, outer wall C_{OW}, window glass C_{WG}, ceiling C_{NC}, and partition C_{PW} is the heat capacity of the building member from the boundary between air and wall to the thickness H_T [m]. Then, using the

Table 7.4 Thermal constants of office building materials

Name of item	Symbol	Volumetric specific heat [kJ/m^3 °C]
Plasterboard	C_{PO}	1000
Clear glass	C_{PG}	1900
Sound absorbing board	C_{PC}	26.0
Air	C_{PA}	1.3
Carpet	C_{PF1}	320
P tile for floor	C_{PF2}	16.00

thermal constants of the materials shown in Table 7.4, it is expressed as Eq. 7.43

$$C_A = C_{PA} \cdot A_R \cdot H_C \tag{7.43}$$

$$C_{NF} = C_{PF1} \cdot A_R \cdot H_{TC} + C_{PF2} \cdot A_R \cdot (H_T - H_{TC}) \tag{7.44}$$

$$C_{OW} = C_{PO} \cdot A_W \cdot H_T \tag{7.45}$$

$$C_{WG} = C_{PG} \cdot A_G \cdot H_T \tag{7.46}$$

$$C_{NC} = C_{PC} \cdot A_R \cdot H_T \tag{7.47}$$

$$C_{PW} = C_{PO} \cdot A_P \cdot H_T \tag{7.48}$$

H_{TC} is the carpet thickness [m]. Therefore, the total heat capacity C_{Hi} subject to air-conditioning is becomes as Eq. 7.49.

$$C_{Hi} = C_A + C_{NF} + C_{OW} + C_{WG} + C_{NC} + C_{PW} \tag{7.49}$$

(2) **Calculation of external intrusion heat load Q_O**

Now, the intrusion heat from the outside of the room is calculated by Eq. 7.50.

$$Q_O(t) = Q_{OW}(t) + Q_{OC}(t) + Q_{OF}(t) + Q_{OG}(t) \tag{7.50}$$

$Q_{OW}(t)$ is the penetration heat from the outer wall [kW], $Q_{OC}(t)$ is the penetration heat from the ceiling [kW], $Q_{OF}(t)$ is the penetration heat from the floor [kW], $Q_{OG}(t)$ is the window glass It is intrusion heat [kW]. The reason for dividing by 1000 is to convert the unit from [W] to [kW].

At this time, intrusion heat from the outer wall considers not only the outside temperature but also the influence of solar radiation. Solar radiation is absorbed by the surface and travels through the wall and becomes a heat load. The effect of solar radiation is converted to an equivalent temperature, and the temperature expressed as if the outside air temperature has risen is called the equivalent outside air temperature *SAT* (Sol Air Temperature) [°C] [7.5].

Furthermore, the intruding heat from the outer wall becomes a thermal load in the room with a time delay due to the heat capacity of the wall, so Eq. 7.51 is used: *ETD* (Equivalent Temperature Difference) [°C].

$$ETD(t) = \frac{\sum_{n=0}^{23} \phi_n \cdot SAT_{h-n}}{K_W} - T_{AIRi}(t) \tag{7.51}$$

where h is the current time (0–23) [h], ϕ_n is the through-wall thermal response coefficient [W/(m^2 °C)], and K_W is the heat transmittance [W/(m^2 °C)] And has Eq. 7.52.

$$K_W = \sum_{n=0}^{23} \phi_n \tag{7.52}$$

From the above, the once-through heat load $Q_{OW}(t)$ [kW] is obtained by Eq. 7.53.

$$Q_{OW}(t) = K_W A_W ETD(t) \tag{7.53}$$

$$K_W = \frac{1}{\frac{1}{a_o} + \sum_{e=1}^{N_e} \frac{d_e}{\lambda_e} + \frac{1}{a_1}} \tag{7.54}$$

In Eq. 7.54, a_o is the outer surface heat transfer coefficient [W/m^2 °C], N_e is the number of outer wall members, d_e is the thickness [m] of member number e, and λ_e is the heat conductivity of member e [W/m °C], a_I is the inner surface heat transfer coefficient [W/m^2 °C].

In addition, *ETD(t)* may be calculated by Eq. 7.55 using the effective temperature difference at the representative points, without using the once-through heat response coefficient ϕ_n.

$$ETD(t) = ETD_T(h) + (T_O(t) - T_{OT}(h)) - (T_{Ai}(t) - T_{AT}) \tag{7.55}$$

where $ETD_T(h)$ is the effective temperature difference [°C] obtained from the effective temperature difference table at time h, $T_O(t)$ is the outside temperature, $T_{OT}(h)$ is the standard outside temperature [°C], and T_{AT} is the standard Temperature (summer 26 [°C], intermediate period 24 [°C]). $T_{OT}(h)$ is the value obtained by adding T_{AT} to the value $ETD_{T0}(h)$[°C] of "Wall type 0" in the effective temperature difference table. That is, it is expressed as follows.

$$T_{OT}(h) = ETD_{T0}(h) + T_{AT} \tag{7.56}$$

From the above, the heat of penetration from the wall can be obtained.

Furthermore, in the case of the top floor, it is necessary to consider the intrusion heat $Q_{OC}(t)$ [kW] from the ceiling. In this case, it is calculated by the following formula and added to the heat load.

$$Q_{OC}(t) = K_C \cdot A_R \cdot ETD_C(t) \tag{7.57}$$

Here, K_C is the heat transfer rate [W/(m^2 °C)] of the ceiling, and is obtained in the same way as Eq. 7.44. ETD_C is the ceiling temperature difference and is calculated separately from the wall surface.

On the other hand, in the case of the ground floor, add the thermal load Q_{NF} from the floor. The floor is not affected by solar radiation, but instead has a once-through heat load due to temperature differences. This is expressed by Eq. 7.58.

$$Q_{NF}(t) = K_{IW} \cdot A_R \cdot (T_O(t) - T_{Ai}(t)) \cdot f_{IW} \tag{7.58}$$

Here, K_{IW} is the heat transfer rate of the floor [kW/m^2 °C] and f_{IW} is the temperature difference coefficient between inside and outside (0.0–1.0).

The intrusion heat load from the window glass is obtained by the following formula.

$$Q_{OG} = K_G \cdot A_G \cdot (T_O(t) - T_A(t)) \tag{7.59}$$

Here, K_G is the heat transfer rate of glass [W/(m^2 °C)]. The Eq. 7.59 is the heat transfer due to the difference between the outside temperature and room temperature and the intrusion heat as heat conduction, and the solar radiation load is calculated separately as explained in (3) below.

(3) Calculation of solar radiation load

The thermal load Q_{OR} due to solar radiation that passes through the window glass can be obtained as Eq. 7.60.

$$Q_R(t) = \frac{1}{1000}(I_G(h) SC A_G) \tag{7.60}$$

where SC is the shielding factor (0.0–1.0) of the glass. $I_G(h)$ is the glass window standard solar radiation acquisition [W/m^2], and the value of time h is used from the glass window standard solar radiation acquisition [7.5]. However, this table only describes the summer and clear weather at the representative point, so it is necessary to correct the values if the conditions are different.

(4) Internal heating load calculation Q_I

In this document, assuming general office buildings, the heat generated by people, lighting, and office automation equipment is defined as the internal heating load Q_I.

$$Q_I(t) = (Q_{IH}(t) + Q_{IL}(t) + Q_{IO}(t)) \tag{7.61}$$

Here, Q_{IH} is the heat load [kW] from the occupants, Q_{IL} is the heat load [kW] due to lighting, and Q_{IO} is the heat load [kW] of the OA equipment.

First, the heat load Q_{IH} from personnel in the room is obtained. Human fever has a sensible heat load due to the temperature difference between body temperature and room temperature, and a latent heat load due to sweating and breathing.

$$Q_{IH} = (SH + LH) \cdot N_H(t) \tag{7.62}$$

Here, *SH* is the sensible heat value [kW/person] of the person, and *LH* is the latent heat value [kW/person]. $N_H(t)$ is the number of people in the room [person].

Next, the heat load due to lighting is obtained by the following formula.

$$Q_{IL}(t) = \varepsilon_L E_L A_R \varphi_L(t) \tag{7.63}$$

Here, ε_L is the heating value [kW/kW] per 1 kW of lighting, E_{IL} is the lighting power density [kW/m^2], and $\varphi_L(t)$ is the operating rate of lighting (0.0–1.0).

As heat generation of OA equipment, heat generation $Q_{IP}(t)$ [kW] of personal computer and heat generation $Q_{IC}(t)$ [kW] of other indoor equipment with heat generation are examined, and each heat load is obtained.

$$Q_{IO}(t) = Q_{IP}(t) + Q_{IC}(t) \tag{7.64}$$

$$Q_{IP}(t) = q_{IP} N_{IP} \varphi_{IP}(t) \tag{7.65}$$

$$Q_{IC}(t) = q_{IC} N_{IC} \varphi_{IC}(t) \tag{7.66}$$

Here, $Q_{IP}(t)$ is the heat load from the personal computer [kW], q_{IP} is the heat generation per notebook PC [W/unit], N_{IP} is the number of units, and $\varphi_{IP}(t)$ is the operating rate (0.0–1.0). Similarly, $Q_{IC}(t)$ is the heat generation [kW] of other indoor equipment that generates heat, N_{IC} is the number of units, and $\varphi_{IC}(t)$ is the operating rate (0.0–1.0).

In this way, an example of obtaining a physical model of a building multi-type-conditioning heat load assuming a fictitious office building structure, building materials, outside air and solar radiation was shown.

7.3 Power System Simulation of Virtual Power Plant

7.3.1 Instantaneous Supply and Demand Analysis of Power System

In the previous sections, the building multi-type air-conditioning power limit and the building heat load modeling were described in detail. In this section, the performance of the virtual power plant to the power system is simulated.

At that time, unless the number of buildings is several hundred and the number of air-conditioners is several thousand, the scale will not be effective for system supply and demand balance control. Simplify from the detailed model in the previous section and configure thousands of air-conditioner simulations.

In this section, the power system supply and demand adjustment analysis model is based on the Institute of Electrical Engineers standard model AGC30 [4]. This analysis model is a supply/demand/frequency control simulation model developed by the Institute of Electrical Engineers of Japan, and was developed for the purpose of standardizing the operation of the base system when examining the effects of introducing solar power plants and virtual power plants. It is a thing. The simulation model configuration is shown in Fig. 7.12.

Table 7.5 shows the AGC30 specification table. On Matlab/Simulink, the AGC30 model is used at the central power supply command center in Japan based on the theory of dynamic characteristics analysis of the load frequency of commercial large synchronous generators. It consists of LFC control and EDC control models, general

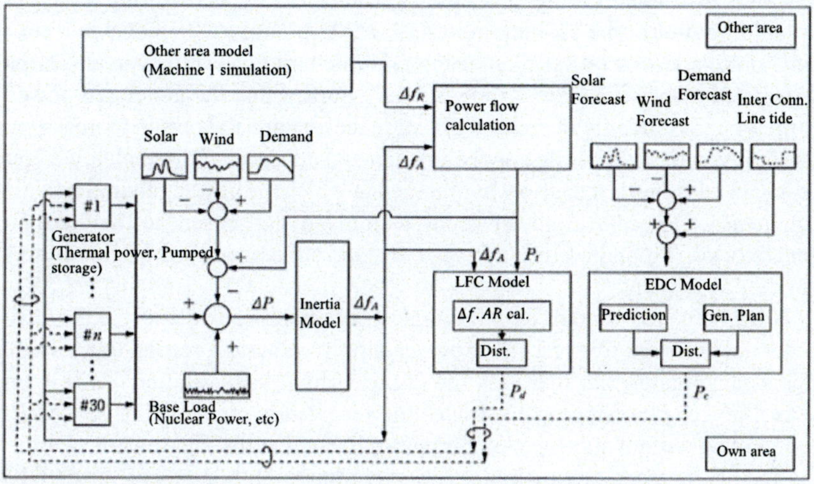

Fig. 7.12 Configuration of AGC30 model (The IEEJ Technical Report: Standard Analysis Model for Power Supply/Demand and Frequency Simulation (2016), p. 23)

Table 7.5 AGC30 model specifications (The Institute of Electrical Engineers of Japan: Power and Energy, quoted from www.iee.jp/pes/?page_id=2009)

Item		Range
Analysis time		About 3 h
Target frequency range		Fundamental frequency ± 0.2 Hz (excluding abnormal situations)
Analysis scale		2 areas (30 target areas, 1 other area)
Standard model	Generator LFC EDC System model	Dynamic characteristic model Thermal machine (steam type fuel) Hydro machine (constant speed, variable speed) FFC, simulate two TBC Output change rate distribution Equilty distribution Inertia model Synchronization coefficient model
Standard data	Demand Distributed power output	Simulate with time series data 2 patterns (Heavy load weekdays, Light load holiday) Simulate with time series data Solar power generation 24 h ± 2 h 5patterns Wind-power generation 6 h ± 2 h 6patterns

power demand fluctuation model and others. Large commercial thermal power generators are modeled for each type of fuel because the responsiveness to supply and demand adjustment control differs for each type of fuel.

In the following, we will model the mechanism of a virtual power plant that functions as if one existing generator is accelerated by aggregating a large number of building multi-type air-conditioning FastADRs. Figure 7.13 shows a conceptual diagram of a virtual power plant using multiple building multi-type air-conditioners distributed over a wide area. Here, it was assumed that the generated power of the entire system suddenly decreased for some reason, and the output increase was saturated only with the existing supply and demand adjustment thermal power generator group. In addition, a scenario was assumed in which the power generation adjustment command of the central power supply command station sent to the thermal power generator corresponding to supply and demand adjustment was originally sent to the virtual power plant.

Now, let us say that an output increase command equivalent to a thermal power generator is issued from a central power supply command station to a virtual power station aggregator that manages the entire virtual power station. At this time, the virtual power plant aggregator issues the received command as a FastADR power suppression command $L(t)$ to the subordinate facility aggregator. Furthermore, the facility aggregator distributes the subordinate buildings as a FastADR power suppression command for each building multi-type air-conditioner.

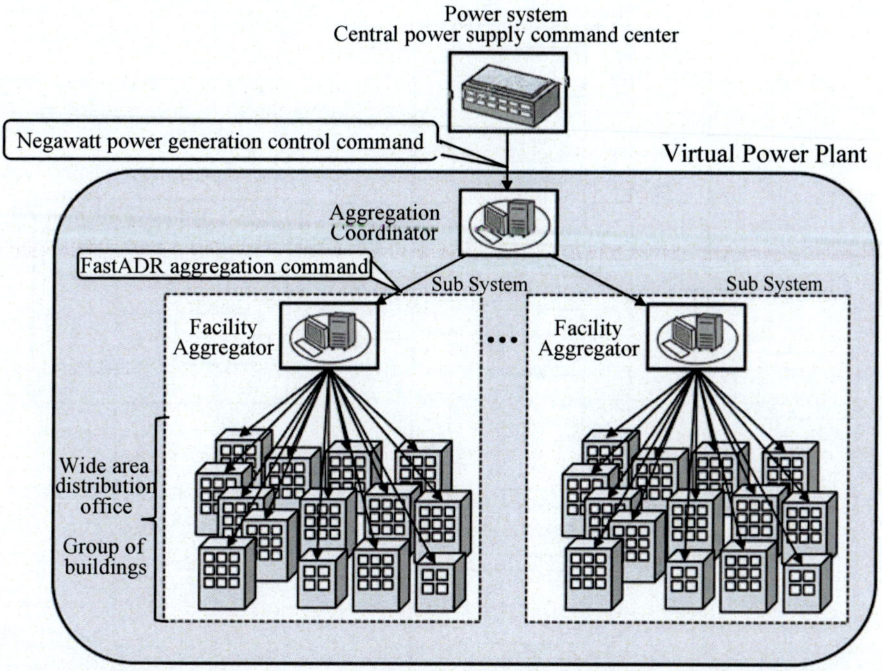

Fig. 7.13 Conceptual diagram of a virtual power plant system with multiple building air conditioners

Therefore, each equipment aggregator and below functions as one subsystem. Each building with multi-type air-conditioning building to be controlled communicates with a higher-level equipment aggregator by a device called ERC (Energy Resource Controller). Each air-conditioner is controlled by the received command. In this chapter, we assume a system in which N_V is the subsystem with N_C units in each building and N_S buildings in each building.

Figure 7.14 shows a block diagram of the AGC30 power system model with the building multi-type air-conditioning group in this chapter. The total system scale was set at 5 million kW, assuming the scale of Shikoku Electric Power and Hokuriku Electric Power. The existing generators for the general demand of 5 million kW are 2 coal-fired power plants (Coal 1, 2), 2 natural gas-fired power plants (LNG 1, 2), 4 gas turbine combined cycle thermal power plants (GTCC1, 2), 1 nuclear power plant (Nuclear).

The steam plant model used in this chapter is described. A steam power plant consists of elements such as a plant control system, boiler system, governor, and turbine. A block diagram is shown in Fig. 7.15. The steam power plant model is a model that calculates the turbine output (mechanical input to the generator) from the output command (LFC, EDC command) and frequency. In frequency control, the output of a steam power plant is delayed by several tens of seconds due to the influence of the boiler system with respect to changes in the output command. In

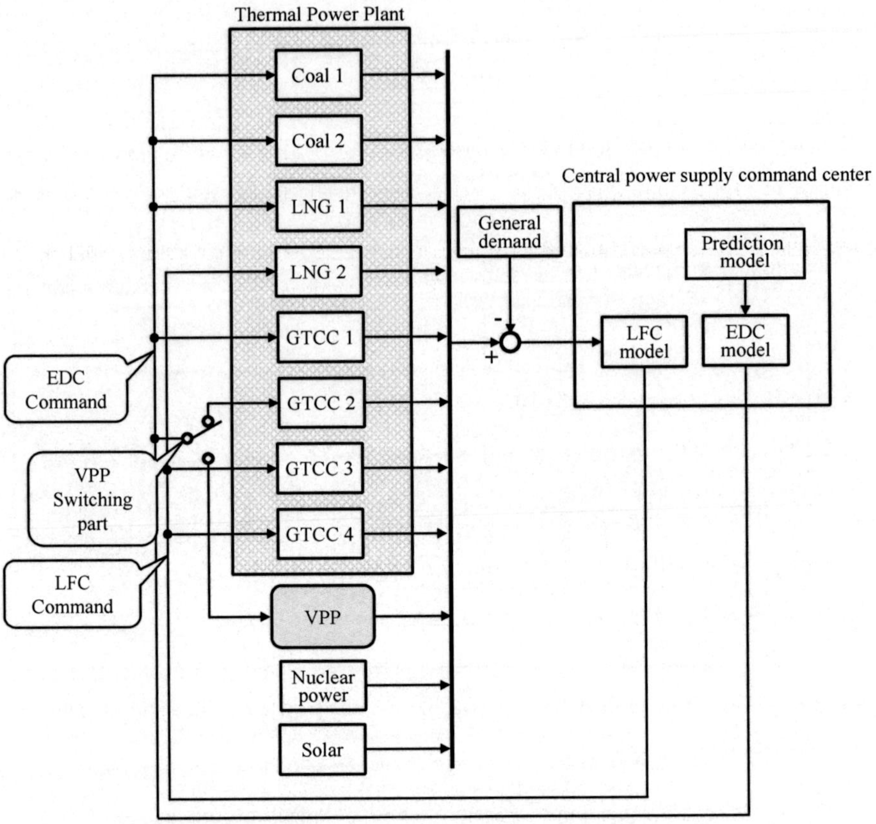

Fig. 7.14 Power system model block diagram including virtual power plant

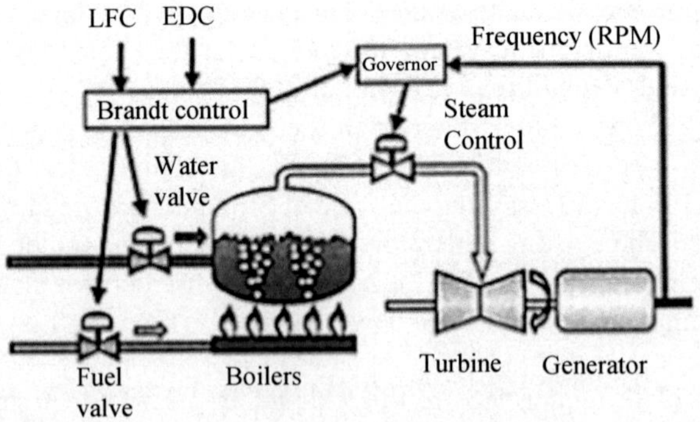

Fig. 7.15 AGC30 steam plant model block diagram (The IEEJ Technical Report: Standard Analysis Model for Electricity Supply/Demand and Frequency Simulation (2016), p. 57)

addition, for relatively long frequency fluctuations, the output is affected by changes in the main steam pressure and load setting control that controls the output from the plant control system to match the command value. Therefore, not only the turbine and the governor, but also the boiler system and the plant control system are modeled as shown in Fig. 7.16. Figure 7.17 shows an overall block diagram of the steam power plant model. General demand and PV (PV) are given as typical time-series data models.

General demand is a light load cross-section model of AGC30. Time series data of demand fluctuations in light load periods where the system capacity is small and the number of generators in parallel is limited based on autumn holidays and it is difficult to secure sufficient frequency adjustment power. Was used. The time series data of general demand is shown in the graph of Fig. 7.18.

On the other hand, the output time series of the photovoltaic power plant uses the AGC30 output decreasing section time series model. Assuming that the solar power generation output decreases greatly between 11:00 and 13:00 when the power

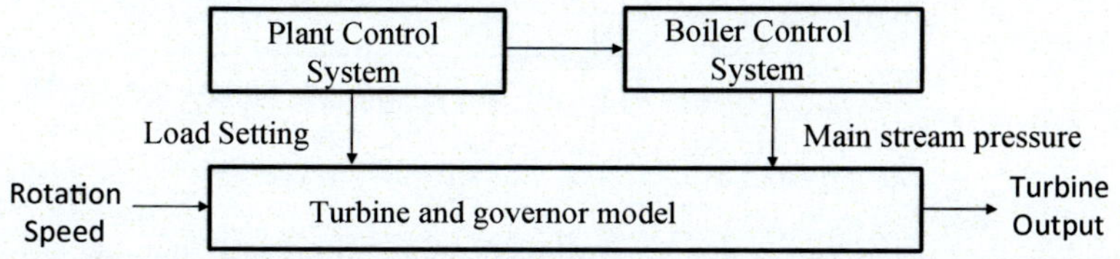

Fig. 7.16 Outline of steam power plant model (The Institute of Electrical Engineers of Japan: Standard analysis model for power supply/demand and frequency simulation (2016), p. 57)

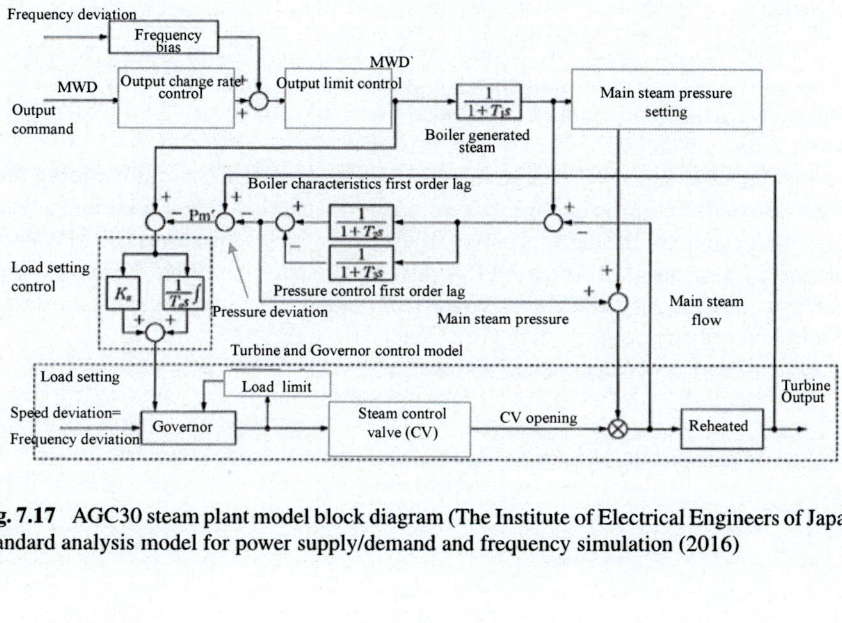

Fig. 7.17 AGC30 steam plant model block diagram (The Institute of Electrical Engineers of Japan: Standard analysis model for power supply/demand and frequency simulation (2016)

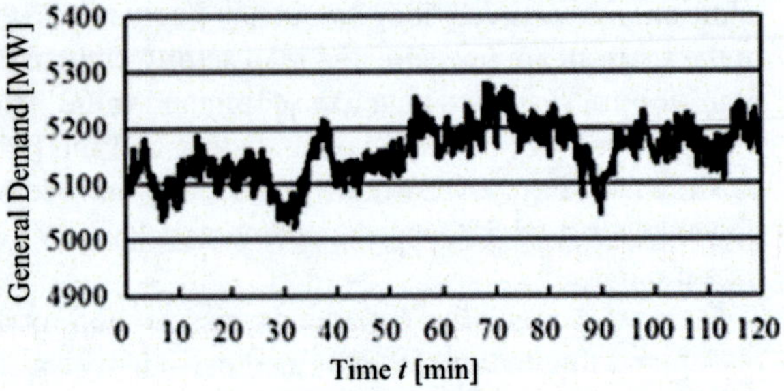

Fig. 7.18 Light load cross section general demand time series graph

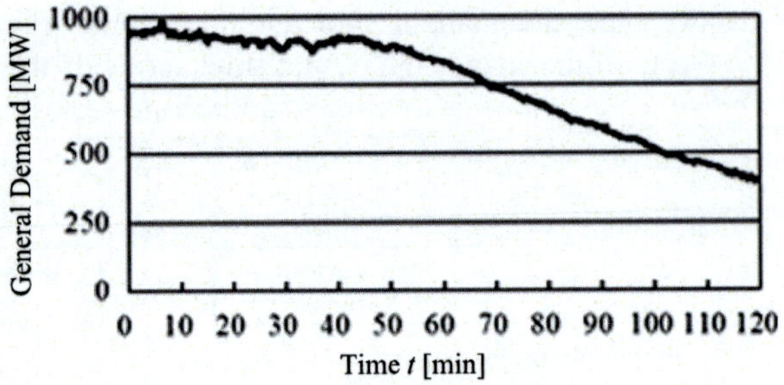

Fig. 7.19 Photovoltaic power generation output time series graph with reduced output cross section

demand is high, assuming that the total time is about 1,000 MW, the power output is constant. Suppose that the state suddenly decreases to about 500 MW in one hour. The output waveform is shown in the graph in Fig. 7.19.

The thermal power plant in charge of supply and demand adjustment is controlled by the central power supply command center in response to this sudden change in solar power output.

In other words, from the central power supply command center, as shown in Fig. 7.14, the LFC (Load Frequency Control) control signal and the EDC (Economic Dispatch Control) signal are commanded every 10 s and every 5 min, respectively, for the adjusted thermal power group. It is done.

In the following, EDC charge thermal power plant CCCC 2 and building multi-type air-conditioning group VPP were added in parallel. Then, assuming that an imminent event that the thermal power adjustment output saturates at 150 MW occurs, we model a scenario in which VPP is also paralleled. Therefore, the central power supply command is a model in which the EDC signal in 5 min increments is originally controlled only by commanding GTCC 2.

7.3.2 Simulation Model of Virtual Power Plant of Building Multi-type Air-Conditioning

Since the compressor inverter of building multi-type air-conditioning power is accelerated and decelerated at a constant rate, the power change with respect to the limit command has a constant rate increase/decrease characteristic. This rate of change has a very different asymmetry between acceleration and deceleration. In order to evaluate the applicability to VPP where high-speed response characteristics are key, it is necessary to model the increase and decrease characteristics of this asymmetric constant rate. Therefore, we use the asymmetric constant rate increase/decrease characteristic model [5] developed in the previous study.

The change in power and room temperature response to the FastADR power suppression limit command is modeled by the following simultaneous recurrence formula.

$$P_{s,c,a}(t+1) = P_{s,c,a}(t) + R_{s,c,a}(t) \tag{7.67}$$

$$\begin{aligned} T_{As,c,a}(t+1) = T_{As,c,a}(t) - k_{ACs,c,a}P_{s,c,a}(t)\cdots \\ + k_{OAs,c,a}\big(T_O(t) - T_{As,c,a}(t)\big) + T_{AEs,c,a} \end{aligned} \tag{7.68}$$

Where t is the discrete time, $P_{s,c,a}(t)$ is the power value [kW] of the air-conditioner a of the subordinate building c of the aggregator s, and $R_{s,c,a}(t)$ is the power increase/decrease Rate [kW/min], $T_{As,c,a}(t)$ is room temperature, and [dig] and $T_O(t)$ are outside temperature [deg]. $k_{ACs,c,a}, k_{OAs,c,a}, T_{AEs,c,a}$ in the room temperature heat balance recurrence Eq. 7.10 This parameter adjusts heat and other factors.

Figure 7.20 shows a conceptual diagram of AE1 operation. t is the discrete time every minute, m is the control frame every 5 min, and $L_{s,c,a}(m)$ is the power limit command value [kW]. The rate of power change $R_{s,c,a}(t)$ is $R_{UPs,c,a}(t)$ [kW/min] during acceleration and $R_{DOWNs,c,a}(t)$ [kW/min]. Even if the power limit value $L_{s,c,a}(m)$ is set to 0, it operates at the minimum compressor speed, so the minimum value is $P_{s,c,a} = P_{MINs,c,a}$ [kW]. The power limit value and the maximum power value are modeled as the rated power $P_{MAXs,c,a}$ [kW].

Furthermore, in order to model the air-conditioning power consumption and room temperature changes, we measured time-series data of actual office air-conditioning air-conditioning power consumption and room temperature every minute and performed parameter fitting. Table 7.6 shows the outline of the actual office building where the measurements were made. Figure 7.21 compares the AE1 model with the actual machine for the power response to the step change of the FastADR power suppression command of the building multi-type air-conditioning system. It was confirmed that the characteristics of the response characteristics, especially the rate of change of the increase and decrease of the power, which are important for the characteristics of the virtual power plant, can be modeled sufficiently.

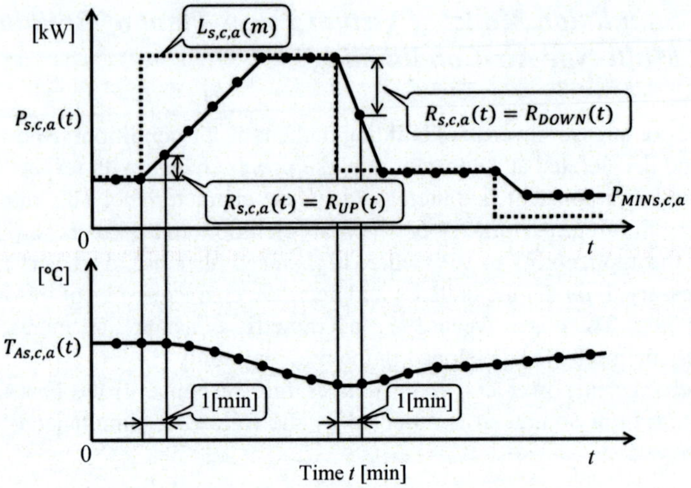

Fig. 7.20 Building multi air conditioner operation conceptual diagram

Table 7.6 Office buildings subject to actual measurement

Architectural form	General office building
Size	2 stories, Total floor area 1600 [m^2]
Number of air -conditioners	5 units
Rated cooling capacity	45, 40, 68, 73, 45 [kW]

However, in some cases, the actual machine time series 16:20–16:25 in Fig. 7.21a should be operated at the minimum speed. Since it is maintenance control and is a transient exceptional operation, it is excluded from modeling.

Table 7.7 shows the parameters of the five AE1 models used here. The parameter is a value that is fitted to the time-series data of measured values per minute of power consumption and room temperature. Here, since the building was equipped with 5 types of multi-type air-conditioning systems, 5 types of different AE1 models were fitted to the actual equipment. Although it is ideal to create more models for rigorous modeling of large-scale building multi-type air-conditioning groups, the FastADR response acceleration/deceleration characteristics are generally the same, and the rest is due to scaling issues determined by the rated capacity of the air-conditioner. Since there are 5 types for each building, different types of response scales are modeled by installing them in each building.

Next, the AE1 model group described in the previous section is installed in the building multi-type air-conditioning group FastADR test facility. Also, in this test facility, a real-time communication model that simulates the delay caused by each communication and negotiation will be constructed by using a different terminal for each aggregator.

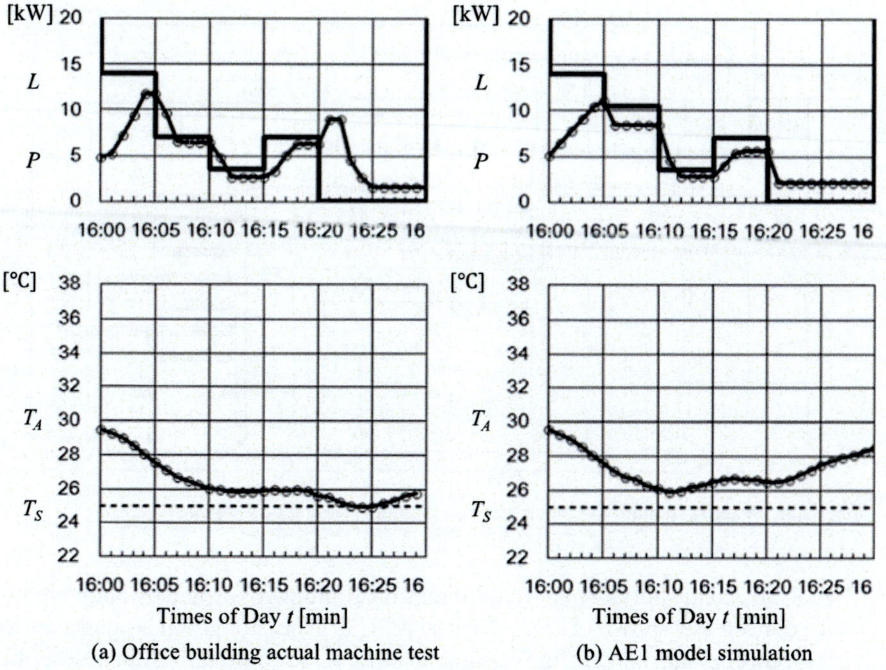

Fig. 7.21 Operation comparison between actual machine and AE1 air conditioner model

Table 7.7 Air-conditioner model variations

Parameter	AC1	AC2	AC3	AC4	AC5
R_{UP}[kW/min]	2.4	1.3	1.8	2.4	1.3
R_{DOWN}[kW/min]	−3.8	−4.2	−5.4	−4.6	−4.5
P_{MAX}[kW]	21	19	25	20	21
P_{MIN}[kW]	2.9	2.2	3.5	4.0	2.9
k_{OA}[deg/min]	0.024	0.087	0.012	0.012	0.012
k_{AC}[deg/(kW min)]	0.029	0.065	0.010	0.014	0.029
T_{AE}[deg/min]	0.060	−0.24	0.015	0.012	0.030

Figure 7.22 shows the communication sequence of the implemented virtual power plant model. In other words, this system consists of a virtual power plant aggregator, equipment aggregator, and AE1 model in the office building. The number of facility aggregator subsystems subordinate to the virtual power plant aggregator is N_V, and the number of subordinate buildings in each facility aggregator subsystem is N_S. Also, in order to give diversity to each building, it is assumed that there are 5 types of AE1 models with different numbers of N_C units, and there are a total of 325 units in one subsystem.

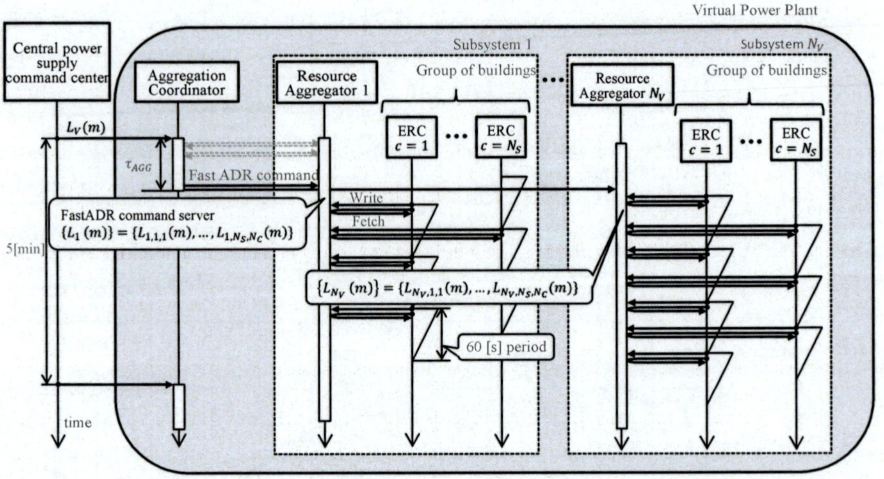

Fig. 7.22 Power system model block diagram including virtual power plant

AGC30 and the building air-conditioning VPP simulator owned by our laboratory were connected as shown in Fig. 7.23. The AGC30 generator group is connected as a virtual power plant via real-time communication. A real-time algorithm is described.

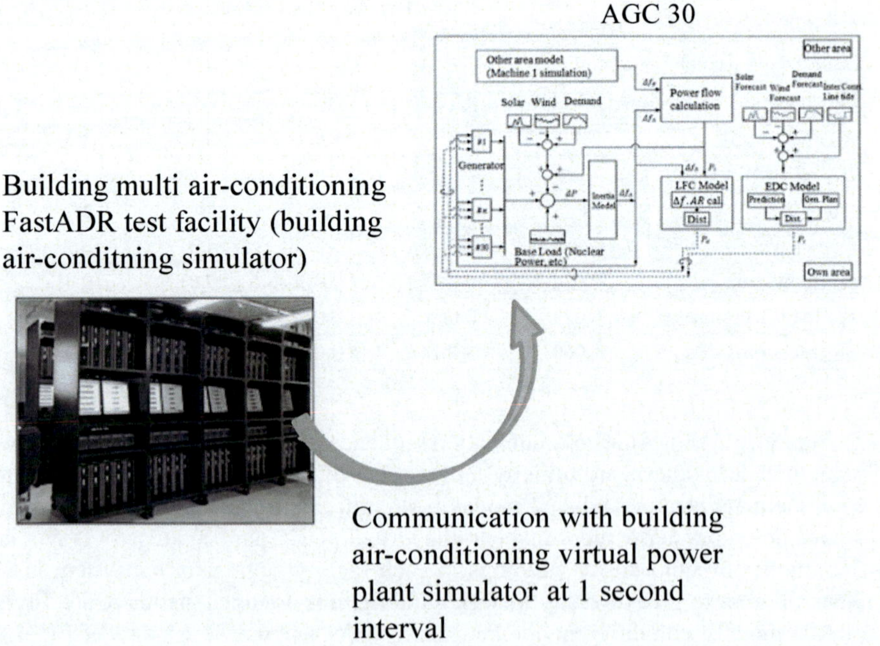

Fig. 7.23 Connection between AGC30 and Simulator

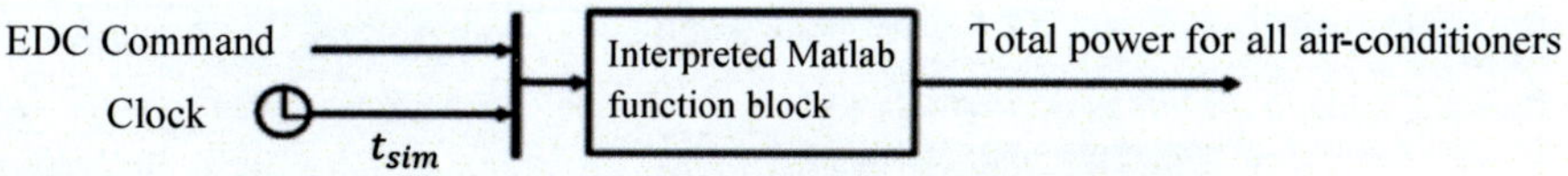

Fig. 7.24 Interpreted Matlab Function block diagram for real-time communication

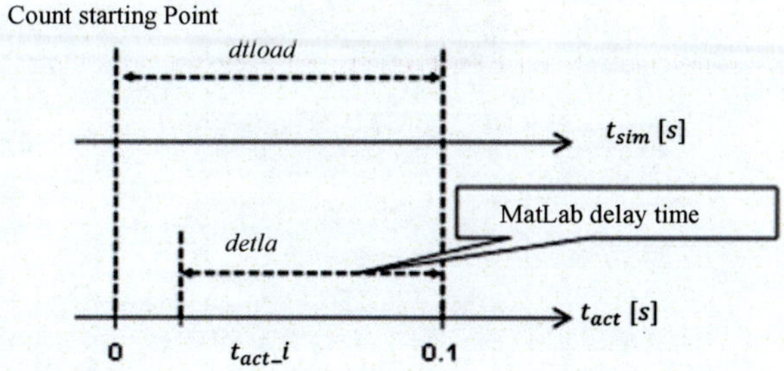

Fig. 7.25 Matlab/Simulink real-time execution timing chart

The Interpreted Matlab Function block shown in Fig. 7.24 is a Simulink block that executes the specified Matlab function.

Figure 7.25 shows a timing chart for synchronous execution with real time. In the Matlab function, the simulation time t_{sim} [sec] and the actual elapsed time t_{act} [sec] measurement by the PC internal clock are started simultaneously with the simulation start. The simulation step *dtload* [sec] of the AGC30 model is 0.1 s. Therefore, the value of t_{act_i} is obtained when t_{sim} is a multiple of 0.1, and the difference *delta* [sec] is calculated.

$$delta = 0.1n - t_{act} \quad (n = 1, 2, \ldots) \tag{7.69}$$

Since delta shows the difference between the simulation time and the actual elapsed time, the simulation time and the real time are synchronized by giving the waiting time to *delta* seconds Matlab/Simulink.

In this way, FastADR real-time simulation with a rated power consumption of 130,000 kW in the entire VPP system will be implemented. The FastADR total suppression command $L_V(m)$ [kW] from the grid is equally distributed to the air-conditioning of each facility aggregator as the power restriction command $L_{s,c,a}(m)$ [kW]. The communication for distribution of restriction commands between aggregators is implemented in real time by actually implementing the international standard OpenADR communication dedicated to demand response in the simulator. In addition to the communication delay, the time delay due to the limit command allocation calculation and the limit value allowance confirmation negotiation is assumed to be uniform τ_{AGG} 60 [s].

The communication from the equipment aggregator to the ERC controlled by FastADR in each building is implemented to actually operate the communication standard IEEE1888 specializing in building equipment remote control.

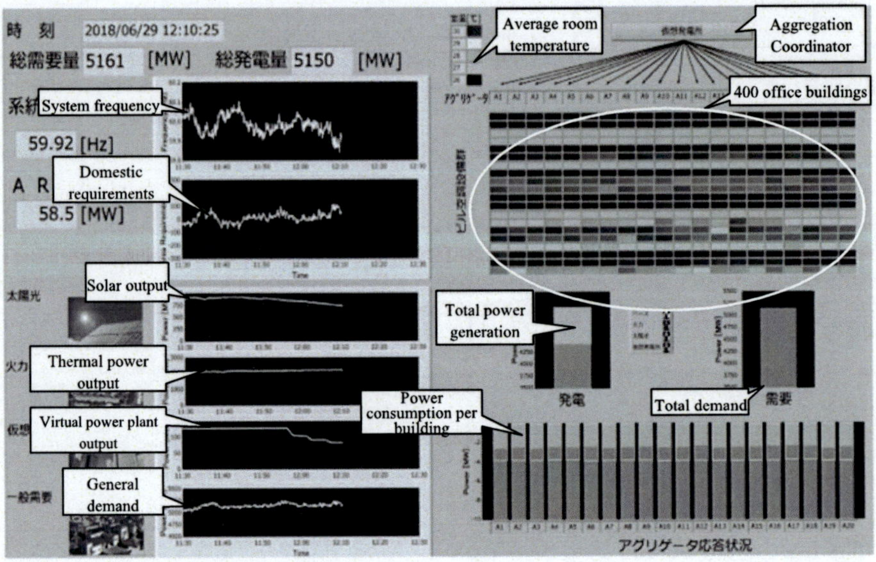

Fig. 7.26 Toy multi monitor of building multi air conditioner virtual power plant

This equipment aggregator sets the power limit command set $\{L_{s,c,a}(m)\}$ for each air-conditioner in its IEEE1888 server (Storage), and each ERC is stored at a fixed period of 60 [s]. The above $\{L_{s,c,a}(m)\}$ Fetch software is implemented and operated. In actual operation, it is assumed that the fixed-cycle Fetch timing of each ERC is probabilistically different, and is uniformly distributed between 0 and 60 [s].

In an actual power system, the status of the system is displayed on the monitor at the central power supply command center. In the simulation, it is useful to be able to grasp the status of the grid and VPP as needed. Therefore, a visualization tool simulating the central power supply command monitor shown in Fig. 7.26 was prepared.

The display item was the monitor of the central power supply command station, but in this document, the virtual power plant by the building multi-type air-conditioning group FastADR is the target of simulation, so the status of the virtual power plant is also the target of the display item. Display items include total demand, total power generation, frequency, AR (Area Requirement), photovoltaic power generation and total thermal power generation, FastADR, average room temperature and aggregator response for each building The situation. With this monitor, you can check the status of FastADR aggregation in each building in real time as well as the supply and demand adjustment results for the entire grid.

In the lower right corner of the screen, how much power is actually suppressed by each subsystem is displayed for each aggregator. The square frame on the bar graph is the sum of the suppression commands for the entire subsystem sent every 5 min. In other words, the building group belonging to one facility aggregator tries to suppress the target amount as a whole, but the upper part of the bar graph represented by two colors actually reduces the electric energy. The total value of all subsystems is displayed at the top of the power generation bar graph as the amount of power generation as a virtual power plant.

7.3.3 Example of Virtual Power Plant Performance Simulation

Figure 7.27 shows the simulation results of whether or not the virtual power plant in the building multi-type air-conditioning group has the same output control responsiveness as the thermal power generator when the output of solar power generation decreases rapidly. A simulation experiment is underway to control the virtual power station by the output control command of the central power supply command station that is issued every 5 min for 30 min after 55 min. In addition, instead of using a virtual power plant for a decrease in solar power generation output, a case where a thermal power generator is used as an alternative to Case A, one GTCC thermal power generator, the virtual power Case B is controlled by switching to a location.

The response of frequency, solar power output, GTCC power output, and total air-conditioning group power consumption is shown in Fig. 7.27. The control period of 30 min ($t = 55$–85 [min]) is indicated by arrows on Fig. 7.27. Case A is shown as a gray line and Case B is shown as a black line. Table 7.8 shows the parameters of the simulation system.

Whether the supply and demand adjustment is good by replacing the virtual power plant in the building multi-type air-conditioning group is evaluated by comparing the frequency error. In fact, looking at the frequency fluctuation graph of the top graph in Fig. 7.27, the frequency time series graphs in Cases A and B match, and there is no difference in the effect on frequency variation.

In order to quantitatively compare the effects on frequency, the 95th percentile value $\Delta Hz_{95\%}$[Hz] of the frequency deviation in the virtual power plant control period of 30 min from 55 to 85 min from the start of the simulation was compared. The results are shown in Fig. 7.28. In case A with only GTCC thermal machine, $\Delta Hz_{95\%\_A} = 0.09$ [Hz], and in case B where the thermal machine is saturated due to a malfunction and replaced with a virtual power station, $\Delta Hz_{95\%\_B} = 0.09$ [Hz], there was no difference, and an example of the simulation experiment result that the response acceleration performance to the output change control command every 5 min from the central power supply command station is equivalent.

In other words, in this simulation, the building multi-type air-conditioning group virtual power plant has the same acceleration response as one GTCC thermal power generator. However, there are further issues as follows.

Here, a simulation experiment was conducted while monitoring the average room temperature deviation for each building by monitoring the entire virtual power plant, but this was a complicated experiment to quantitatively distribute control commands

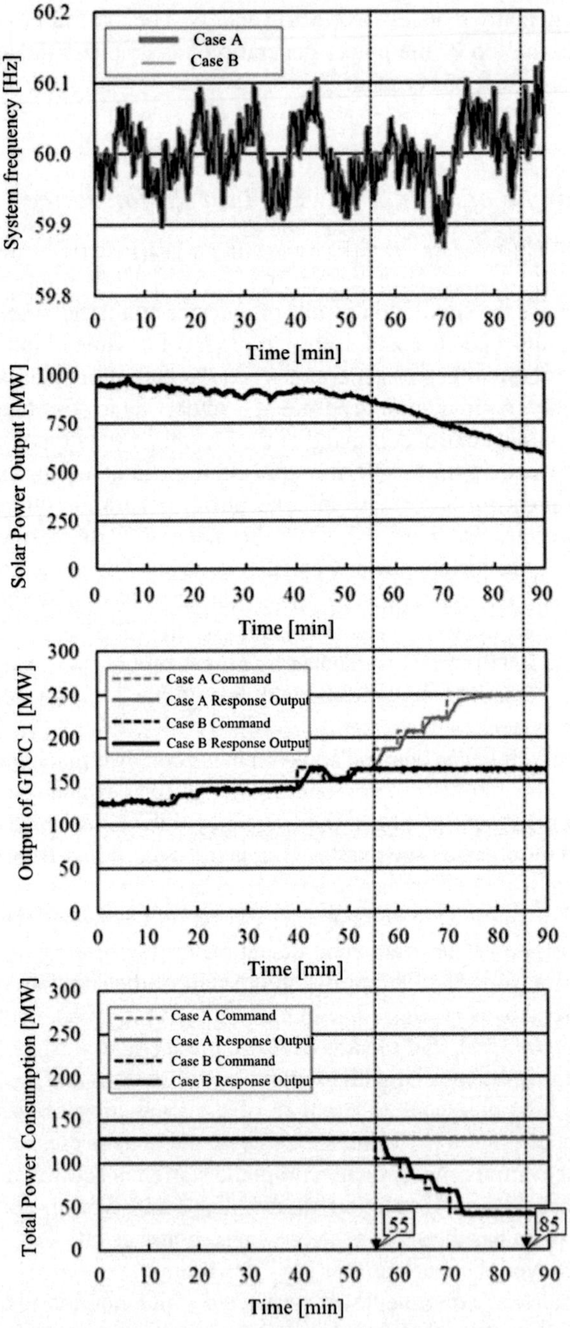

Fig. 7.27 Result of virtual power plant frequency control simulation

Table 7.8 Power system simulation model parameters

Parameters	Value
System constant	10 [%MW/Hz]
Load frequency characteristics	2 [%MW/%Hz]
Composite inertia constant	49.0 [s]
LFC PI control proportional gain	1
LFC PI Control integral gain	0.003
LFC PI control cycle	5 [s]
EDC sampling time	300 [s]
Reference frequency	60 [Hz]
GF capacity	3 [%]
LFC capacity	2 [%]
Own area driving reserve	15 [%]
Interconnected power flow value	0 [MW]
	2.5 [s]
Interconnected tidal flow fluctuation cycle	TBC
Frequency control method	60 [s]
Negotiation time τ_{AGG}	20
Number of subsystems N_V	20
Number of buildings per subsystem NS	4–20
Number of air conditioners per building NC	325

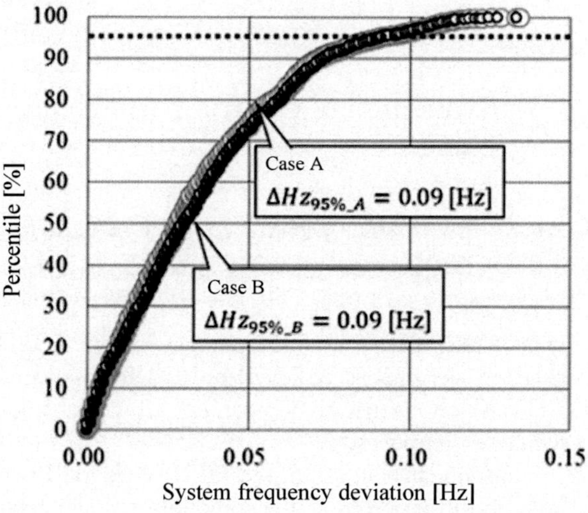

Fig. 7.28 Grid frequency deviation percentile

for the factor of maintaining room temperature comfort, which is a side effect of FastADR.

In addition, when a thermal power plant is switched to a virtual power plant due to an event where saturation acceleration cannot be performed due to some trouble with respect to the acceleration command of the central power supply command station, as an alternative scenario, the output does not come out as planned, so the load frequency control deteriorates. Although it was evaluated as frequency control as an index to be able to complement it, the economic effect was not evaluated.

The advantage of substituting a thermal power plant with a virtual power plant is not limited to emergency replacement when such a malfunction occurs. Virtual power plant substitution may be able to improve such events as the annual operating frequency is very low, the economic burden of maintaining a standby thermal power plant, and the standby CO2 emissions of immediate thermal power.

In this chapter, the effect on the frequency control performance in the case of a large-scale integrated virtual power plant, which has been a problem with building multi-type air-conditioning power FastADR, was simulated using a power system real-time simulation facility. As a result, the response to the output increase command of the virtual power plant by the building multi-type air-conditioning group was the same as that of one thermal power generator.

This result may be utilized when considering how much control performance the building air-conditioning FastADR has when compared to conventional generators. However, there are issues such as how much control performance the negotiation time of the virtual power plant aggregator has under actual feasible conditions, and how the response rate to the command changes when considering room temperature. This topic will be dealt with in the following sections.

7.4 Evaluation of Control Performance of Virtual Power Plant

7.4.1 Example of Evaluation Criteria for Grid Operation Organizations in United States

In the United States, separation of power generation and transmission is progressing. In 1992, the energy policy stipulated Independent Power Producer (IPP), which allowed non-electricity companies to own, operate and sell power generation for wholesale purposes.

Subsequently, in 1996, the FERC (Federal Energy Regulatory Commission) in order to further promote the competition in the power wholesale market, in order to wipe out concerns about the preferential treatment of power generation by each power company. The functions of the power generation and transmission sectors were separated. The separated power transmission department became an ISO (independent grid operation organization) independent of the power company.

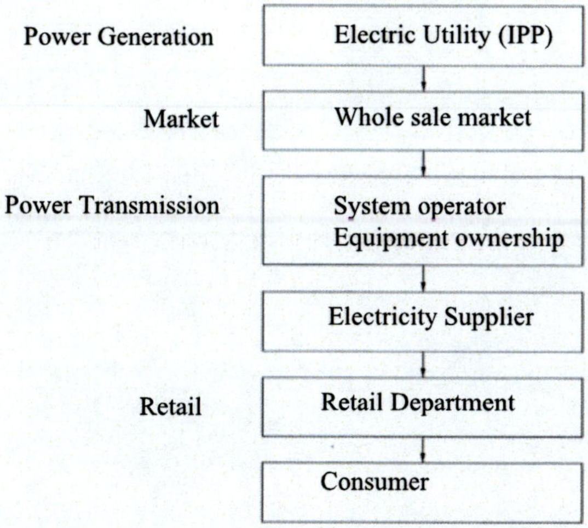

Fig. 7.29 US electricity supply system diagram

The ISO keeps the power grid ownership in the power company, pre-aggregates the IPP power supply plan in the pipe, maintains the balance of power supply and demand, and maintains the frequency in real time. It also operates the electricity wholesale market. In addition, RTO (Regional Transmission Organization) is a wide-ranging ISO across regions.

Figure 7.29 shows the US power supply system. The liberalization of the electric retail market varies from state to state. Electronic retailing is now fully liberalized in 13 of the 50 states and Washington, DC. In particular, the state of Texas has an environment in which free competition between retailers is facilitated, the retail market is most successful, and almost all consumers are switching to new businesses.

On the other hand, in order to liberalize the electric power market in California, an electric company sold power generation facilities and decided to thoroughly distribute electricity. However, the sale of power generation facilities would lead to a lack of power generation capacity, and fossil fuel Due to the soaring price, the wholesale price of electricity has exploded. Electricity companies are unable to raise their electricity bills due to regulations, making it difficult to wholesale electricity to consumers. Finally, in the summer of 2000, due to the tight demand caused by extreme heat, power supply is insufficient and power outages I fell into a situation. Therefore, in September 2001, the state of California canceled retail liberalization.

On the other hand, ISO/RTO currently manages a power grid connected to a large number of power generation companies, Fig. 7.30 shows each area. Large-scale ISO/RTO in the US includes CAISO (California ISO), ERCOT (Electric Reliability Council of Texas), PJM Interconnection, Southwest Power Pool, NYISO (New York ISO), ISO-NE (New England ISO), MISO (Midcontinent ISO). PJM has established a negawatt trading market, sending balancing signals (frequency control

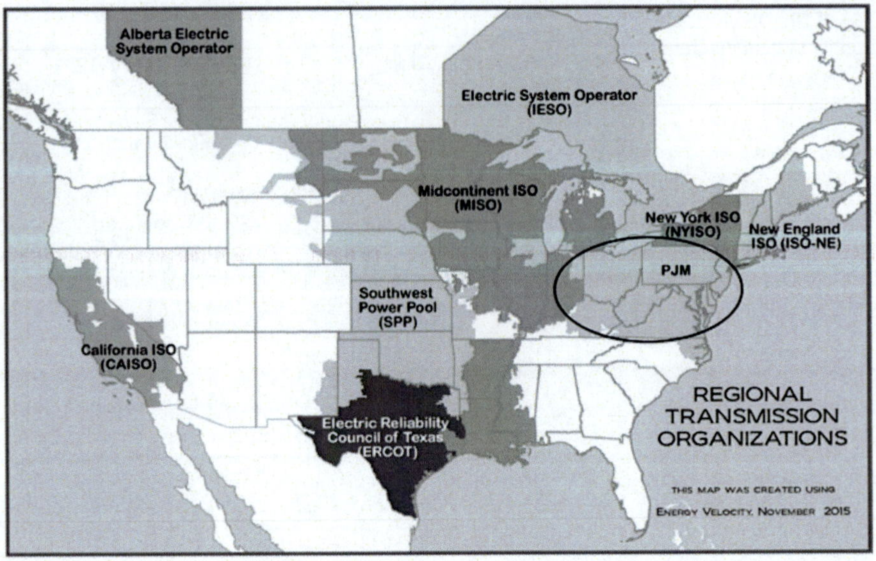

Fig. 7.30 US ISO/RTO (FERC: Regional Transmission Organization (RTO)/Independent System Operator (ISO)

signals) to consumers with power generation facilities and negative wattage equipment, and performing frequency control by negawatt trading. In order to realize ancillary services such as load frequency control of power grids using negawatts, it is necessary for consumers to follow the balancing signals for negawatt from the grid side in minute and minute increments.

Among them, PJM located in the eastern United States has the best smart grid technology. PJM has established a negawatt trading market, sending balancing signals (frequency control signals) to consumers with power generation facilities and negative wattage equipment, and performing frequency control by negawatt trading. Thus, in order to realize ancillary services such as load frequency control of the power system with negawatt, consumers who follow the balancing signal for the negawatt from the system side precisely and in increments are required.

7.4.2 Building Multi-type Air-Conditioning Virtual Power Generation Control Model

A conceptual diagram of the building multi-type air-conditioning FastADR system assumed in this document is shown in Fig. 7.31. The evaluation target is a system that receives a balancing signal from the power system operator, distributes the FastADR power suppression value from the intermediary service provider (aggregator), and

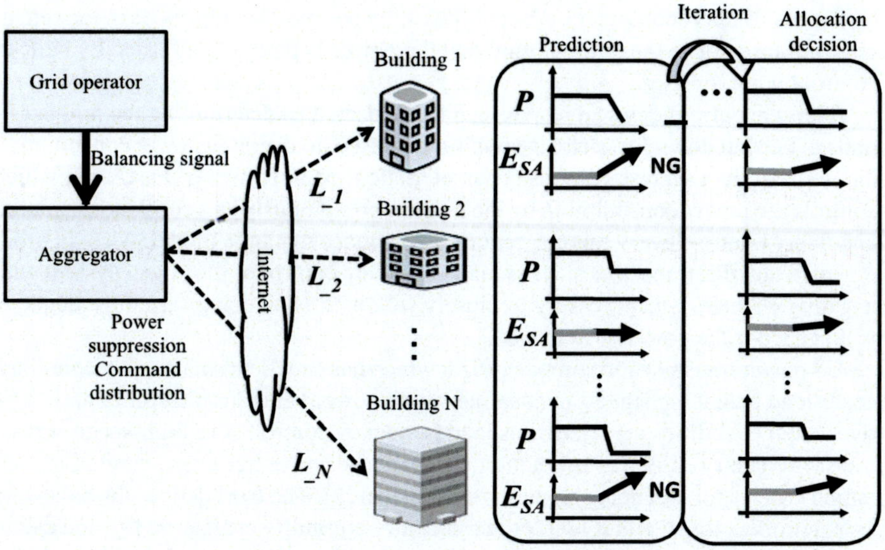

Fig. 7.31 Allocation of FastADR power control command to building multi air conditioners in buildings

aggregates and adjusts the negative wattage response from the building multi-type air-conditioning system in the N building.

In this section, the virtual power plant control performance is shown only by FastADR aggregation of the building multi-type air-conditioning group, so the compensation control combined with the power storage battery described earlier in this document is not included. This is because it is self-evident that the follow-up evaluation passes if the battery is used to follow the balancing signal from the grid.

Here, it is assumed that the aggregator distributes the FastADR power suppression command $L\_$ to subordinate buildings every time it receives a balancing signal from the power system operator. At that time, it is necessary to predict whether the trade-off room temperature rise E_{SA} can be kept within the allowable range while securing the necessary negative wattage response P, and to determine each FastADR power suppression command value $L\_$ [6, 7]. The room temperature rise index for each building is the representative value E_{SA}, which is the average of the deviations between the room temperature per minute (indoor air intake air temperature) and the setting for each building.

For example, as shown in Fig. 7.30, in the first allocation, the trade-off temperature rise E_{SA} is increased in some buildings.

If NG, the command value $L\_$ for the NG building is modified, and the response result P and trade-off room temperature rise E_{SA} are re-predicted and re-allocated.

Building multi-type air-conditioning is the most popular air-conditioning equipment in medium-sized or smaller office buildings. Usually, dozens of indoor units are connected to each outdoor unit by piping, and refrigerant built-in control is performed by dozens of microcomputers. About 90% of the power consumption is in the refrigerant compressor in the outdoor unit, and the power is precisely adjusted by built-in control by the inverter.

As in the current demand response, a method such as determining the number of units to be shut down for each air-conditioner based on the rated power consumption did not require a precise response characteristic model. However, FastADR, which controls the power consumption by the inverter continuously as a continuous amount and finely controls every minute, requires a response dynamic model. Also, in order to realize ancillary services such as load frequency control of the power system with negative wattage, it is necessary to closely follow the balancing signal for negative wattage from the power system side.

US power transmission company PJM, which has excellent smart grid technology, sends load balancing signals to customers with power generation facilities and negative wattage facilities, and performs load frequency control. As a high-speed service transaction entry eligibility evaluation, the response power tracking quantitative evaluation criteria for balancing signals is established [8]. The PJM defines the balancing signal profiles RegA and RegD for market entry eligibility evaluation [9], and calculates and evaluates the negative watt generation for the 40-min balancing signal, that is, the control performance of the power suppression amount as the following Performance Score S_T. Yes. In this document, the low-speed version of RegA was targeted as a balancing signal, and the follow-up of the power reported to the power system operating organization was evaluated. The PJM balancing signal has already been put into practical use, and it has been adopted in the follow-up evaluation of the building air-conditioning power FastADR in references [10, 11].

PJM's balancing signal follow-up evaluation formula is from [8] as Eq. 7.70.

$$S_T \triangleq \frac{1}{3}S_C + \frac{1}{3}S_D + \frac{1}{3}S_P \tag{7.70}$$

Here, S_C is the Correlation Score representing the degree of correlation, S_D is the Delay Score representing the degree of delay, S_P is the Precision Score representing the accuracy, and S_T is calculated from the data for 10 min every 10 s. Since the model used this time outputs data every minute, calculation is performed assuming that the value is held for one minute.

S_C and S_D are evaluated on average over the test time of 40 min by calculating the correlation function with the balancing signal every 5 min by Eq. 7.71.

$$S_C \triangleq \frac{\sum_{i=1}^{I}\left(P_A(i) - \overline{P_A}(i)\right)\left(R_A(i) - \overline{R_A}(i)\right)}{\sqrt{\left(\sum_{i=1}^{I}\left(P_A(i) - \overline{P_A}(i)\right)^2\right)\left(\sum_{i=1}^{I}\left(R_A(i) - \overline{R_A}(i)\right)^2\right)}} \tag{7.71}$$

$i = 1,2, \ldots$, I is the number of data per 10 s in the 5 min window, $P_A(i)$ [kW] is the value that the aggregator collects the power of each air-conditioner and reports to PJM, $\overline{P_A}(i)$ [kW] is the reported average air-conditioning power for 5 min, $R_A(i)$ is the RegA balancing signal value, and $\overline{R_A}(i)$ [kW] is the average.

$$S_D \triangleq \left| \frac{\delta - 300}{300} \right| \tag{7.72}$$

δ is the window start time that maximizes the window S_C value for 5 min.

$$S_P \triangleq 1 - \frac{1}{J}\sum_{j=1}^{J}\left| \frac{P_A(j) - R_A(j)}{\overline{R_A}} \right| \tag{7.73}$$

$j = 1,2, \ldots, J$ is the number of data for 40 min.

In the PJM standard, the entry qualification is passed if the Performance Score S_T in Eq. 7.70 is 0.75 or more [8]. PJM distributes the Score calculation tool [12] based on the above Eqs. (7.70–7.73), but since parameters that cannot be obtained without actually contracting are added to the calculation, The score is 10% different from the above calculation. Therefore, in this book, without using a tool, the calculation was judged directly from Eqs. (7.70 - 7.73).

Figure 7.32 shows the aggregation process model in the simulation. The RegA signal from the power system operating organization to the aggregator is a dedicated communication protocol such as OpenADR [13], and reads out the RegA from the aggregator every minute. The simulation conditions are the simulation model of the aggregator as shown in Table 7.9. In this example, N = 10 buildings in order to match the minimum load scale for PJM balancing evaluation. However, since this model is a linear model, it is scalable in the following superposition and can be simulated in any N building.

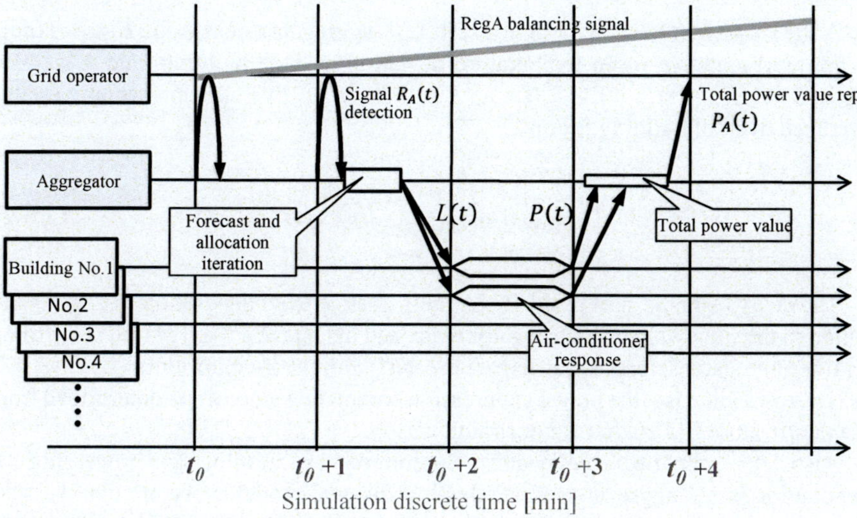

Fig. 7.32 Concept of discrete time simulation model of aggregation processing time sequence

Table 7.9 Simulation conditions

Parameters	Value
Total number of buildings N	10
Total Negawatts that can be generated	500 [kW]
Rated power per building	150 [kW]
Suppressable power per building	50 [kW]
Number of outdoor units per building	3
Number of indoor units per building	18
RegA signal readout cycle τ_R	0–1 [min]
Aggregation command delay τ_I	1 [min]
Air conditioner response delay τ_A	1 [min]
Aggregation report delay τ_A	0–1 [min]

$$L_A(t) \triangleq \sum_{n=1}^{N} L_n(t) \tag{7.74}$$

Here, $R_A(t)$ is the read value of RegA signal every minute $L_A(t)$ [kW] is the power suppression command value $L(t) = L_1(t), \ldots L_n(t), \ldots L_N(t)$ total $\Delta P_A(t)$[kW] is the negative wattage.

$$\Delta P_A(t) \triangleq \sum_{n=1}^{N} \Delta P_n(t) \tag{7.75}$$

$\Delta P(t) = \Delta P_1(t), \ldots \Delta P_n(t), \ldots \Delta P_N(t)$'s aggregate report value $E_{SAA}(t)$ [deg] is the representative room temperature deviation in each building, and $E_{SA}(t) = E_{SA}(t) = E_{SA\_1}(t), \ldots E_{SA\_n}(t), \ldots E_{SA\_N}(t)$ is an overall representative value averaged over all building groups.

$$E_{SAA}(t) \triangleq \frac{1}{N} \sum_{n=1}^{N} E_{SA\_n}(t) \tag{7.76}$$

The delay from the start of the RegA ramp change from the grid operating organization to the time when the aggregator reads and detects at a fixed period is τ_R[min], and the aggregator selects to participate in each building air-conditioning OptIN/OUT [13] negotiation Also, the power suppression command value $L(t)$ is determined from the temperature rise trade-off constraint.

Also, let τ_L[min] be the delay due to the command value allocation processing and negotiation in the aggregator. Since this is a discrete model every minute, the state quantity $P(t)$ responds with a delay of one step (1 min) from the manipulated variable u (t-1) = L (t-1), and this $L(t)$ The delay in the response of the air-conditioning power τ_P[min] was used. Assume that the aggregator aggregates the air-conditioning power response of each building and reports it to the grid side at regular intervals, and the delay is τ_A[min].

In the calculation of Performance Score S_T, the initial state was set to 0.0 kW, and the followability of the negative wattage change $\Delta P\ (t)$ with respect to the increase/decrease change $\Delta R_A(t)$ of the RegA signal was evaluated. The trade-off comfort was evaluated using the change $E_{SAA}(t)$ from the average room temperature deviation $E_{SAA}(t)$, which represents the entire building group, with the RegA signal start point as the reference 0.0° C. Regarding the tolerance limit of trade-off comfort, referring to previous studies [6, 7], the change from the initial state to $E_{SAA}(t)$ to $\pm$ 0.5 °C is set as an example of the tolerance limit.

In the simulation, it is necessary to try multiple air-conditioning operation conditions in addition to the building multi-device itself. This time, both the internal heat generation load due to the internal activity and the intrusion heat load due to the difference between the outside temperature and room temperature are given conditions as the air-conditioning load from the viewpoint of heat supply and demand. The heat supply and demand conditions are the initial value $L\ (0)$ of the power suppression command, the initial value $L\ (0)$ of the actual power consumption, and the initial value $T_{AA}(0)$. High power condition (i) ($L(0) = 25$ kW, $P(0) = 18$ kW, $T_{AA}(0) =$ 24 °C), medium power condition (ii) ($L(0) = 20\ kW, P(0) = 14\ kW, T_{AA}(0) = 25$ °C), and multiple power supply and demand conditions are set as low output conditions (iii) ($L(0) = 15$ kW, $P(0) = 10$ kW, $T_{AA}(0) = 26$ °C).

7.4.3 Evaluation Example of Building Multi-type Air-Conditioning Virtual Power Generation

Figure 7.33 shows FastADR negative wattage follow-up simulation results of building multi-type air-conditioning power against PJM balancing signal when each processing time delay is $\tau_T = \tau_R + \tau_L + \tau_P + \tau_A = 4$ min.

Figure 7.33a shows the balancing signal absolute value $R_A(t)$, the absolute value $P_A(t)$ of the aggregate building power report command absolute value $L_A(t)$ from the aggregator to the building, and the absolute value $P_A(t)$ of all building negawatt aggregate report values. Each change is shown. Figure 7.32 (b) shows the changes of $\Delta R_A(t)$, $\Delta L_A(t)$, and $\Delta P_A(t)$ per minute from the initial test time period.

If the lamp change $\Delta R_A(t)$ is distributed evenly every minute by dividing $\Delta L_A(t)$ by N evenly in all N buildings, the change in the air-conditioner's characteristic model output will be 1/N times smaller. The response accuracy is not enough. Therefore, in this scale of N = 10, the power suppression command $\Delta L\ (t)$ was added at the same time for every two buildings so that each $\Delta L(t) = 10$ kW.

As shown by the circle in Fig. 7.32b, the change $\Delta P_A(t)$ in the reported value of the negawatt was almost the same as the power change width $\Delta R_A(t)$ of RegA. However, the delay in signal tracking was not improved, so the performance score $S_T = 0.69$, the passing score 0.75 was not reached, and the achievement rate was

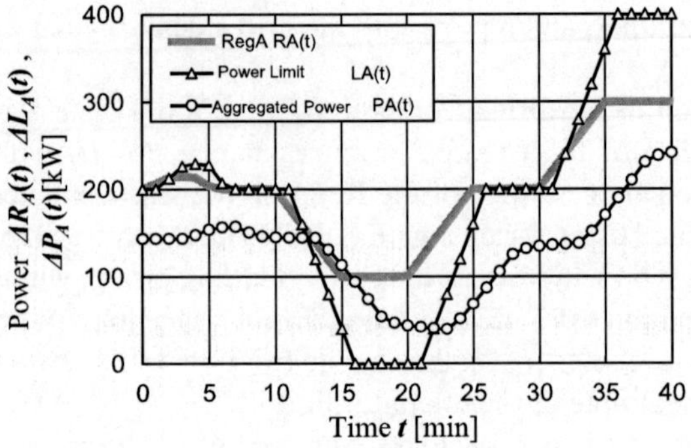

(a) Absolute value of aggregated power $P_A(t)$

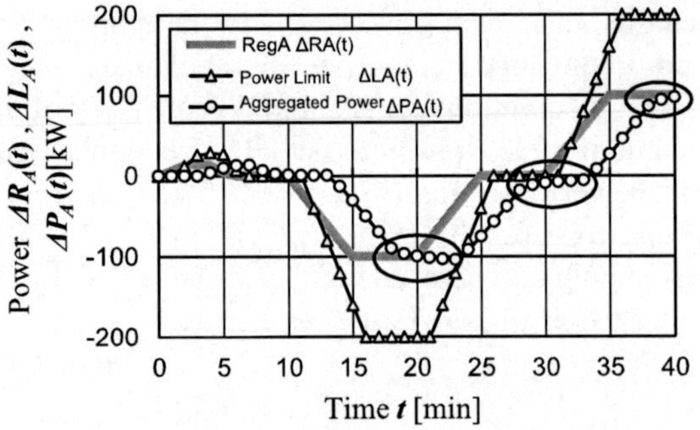

(b) Change in aggregated power $\Delta P_A(t)$

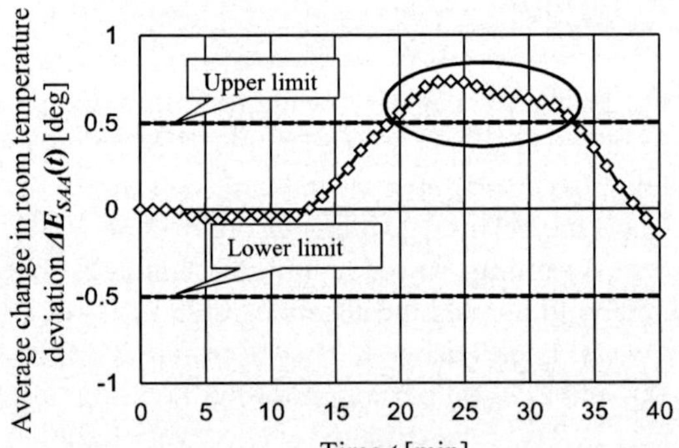

(c) Average change in room temperature deviation ΔE_{SAA}

Fig. 7.33 RegA balancing signal tracking characteristics (communication and processing delay total $\tau_T = 4$ min) only 88%. If $\Delta L(t)$ is changed significantly, the trade-off representative average room temperature deviation $\Delta E_{SAA}(t)$ may deviate from the allowable range. Therefore, it is indispensable to predict and calculate $\Delta E_{SA}(t)$ of each building from the state space model.

In addition, as shown in Fig. 7.32c, the average room temperature deviation deviated from the permissible change + 0.5 deg from around 20 min in this example. The performance score S_T not only reached the passing score by just increasing the power change width $\Delta P_A(t)$, but the side effect of the trade-off room temperature increase was excessive.

The performance score S_T did not pass by just matching the RegA signal change level of the negative watt change level $\Delta P_A(t)$, so this time we tried to improve it by shortening the delay time.

Figure 7.33 shows the case where the total processing delay time can be shortened to $\tau_T = \tau_R + \tau_P =$ 2 min. This is the case when the aggregator performs command distribution instantaneously, and $\Delta L\ (t)$ can be commanded without delay ($\tau_R = 0$) and negative wattage aggregation report can be instantaneously ($\tau_A = 0$). As in the example in Fig. 7.32, the power suppression command amplitude ΔL is an example in which the command is increased by 1.2 times $\Delta R_A(t)$.

With regard to the evaluation score S_T, as shown in Fig. 7.34 (i)–(iii), (i) $S_T =$ 0.76 for high output heat supply and demand conditions, (ii) $S_T = 0.75$ for medium output heat supply and demand conditions, (iii) $S_T = 0.76$ in the low power near heat supply and demand conditions. The negawatt aggregate report change level does not reach the signal change level, but as a result of the delay being greatly improved, the Performance Score S_T is 0.74 to 0.76 under multiple heat supply and demand conditions, and is generally close to the PJM passing score of 0.75.

Regarding trade-off temperature rise, as a result of appropriately selecting the width of ΔL, as shown in Fig. 7.35 (i)-(iii), (i) minimum $\Delta T_{AA}(t) = +\ 0.58$ deg The minimum $\Delta T_{AA}(t) = +\ 0.40$ deg, (ii) medium output heat supply and demand conditions, and (iii) the minimum $\Delta T_{AA}(t) = -0.55$ deg deg at low output heat supply and demand conditions. $\Delta E_{SAA}(t)$ was + 0.58 to −0.55 deg throughout the test time, and it was possible to examine in detail how much it deviates from the target ± 0.5 deg.

The PJM RegA balancing signal changes in a 10-s cycle, but this model was read in a 1-min cycle, which is the current market performance communication cycle, so the delay was large and tracking was difficult. Another reason for the delay is the discrete state space representation model every minute, and the state quantity $\Delta P\ (t)$ is delayed by one step 1 min from the manipulated variable $u(t-1) = \Delta P_L\ (t-1)$. It was because it responded. Although the reason for this delay is a model restriction, the delay time is almost the same as the delay until the building multi-embedded control accelerates/decelerates the compressor inverter from the BEMS via the communication interface.

In order to satisfy market entry qualification, it is necessary to shorten the time required to determine the power suppression command value for each building and to aggregate the response power of each building by various previous studies [14–17]. It is. Since the state-space model in this document can express not only the power

a

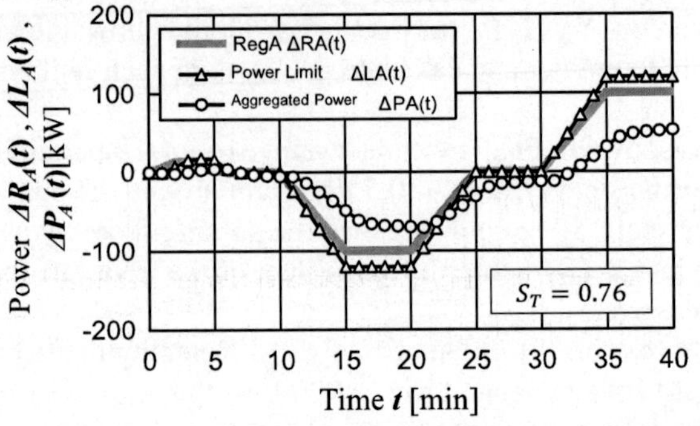

(i) L(0)=25kW, P(0)=18kW, TAA(0)=24deg

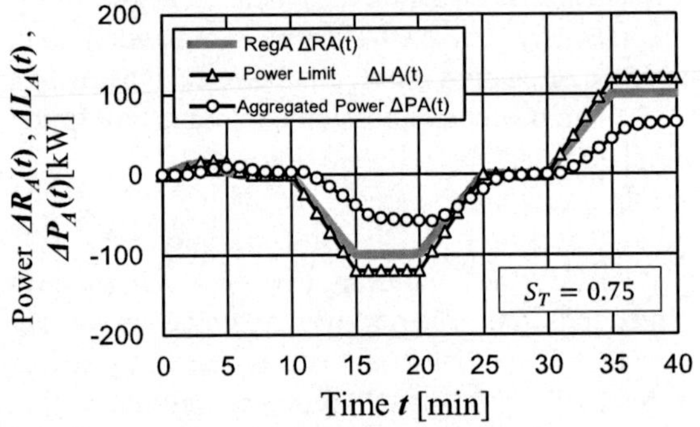

(ii) L(0)=20kW, P(0)=14kW, TAA(0)=25deg

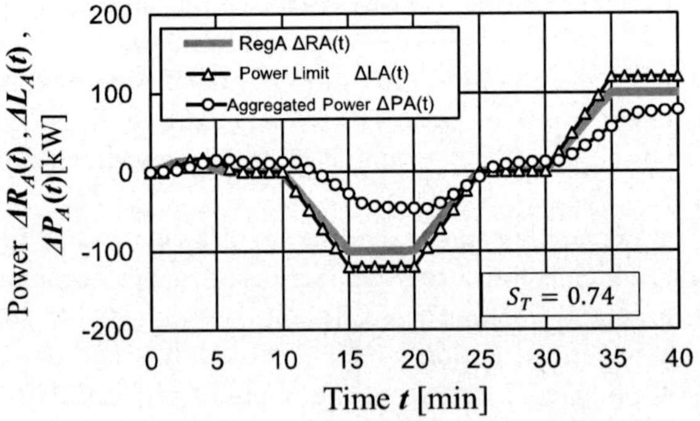

(iii) L(0)=15kW, P(0)=10kW, TAA(0)=26deg

Fig. 7.34 Change in aggregated power $\Delta P_A(t)$ for RegA balancing signal tracking characteristics (communication and processing delay total $\tau_T = 2$ min)

b

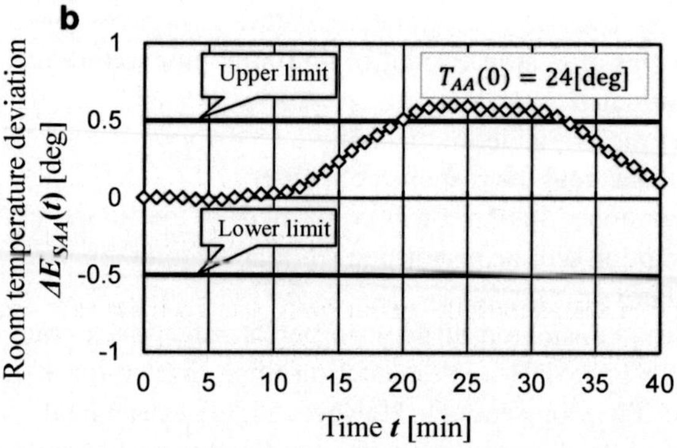

(i) L(0)=25kW, P(0)=18kW, TAA(0)=24deg

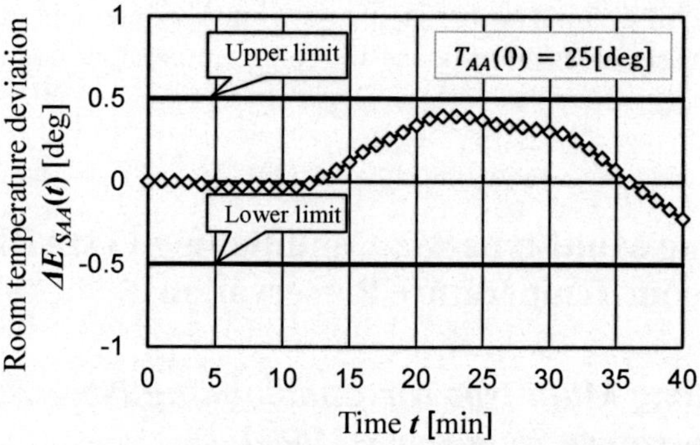

(ii) L(0)=20kW, P(0)=14kW, TAA(0)=25deg

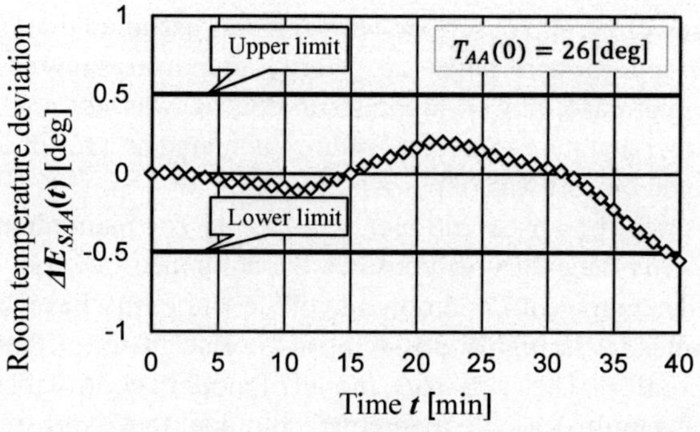

(iii) L(0)=15kW, P(0)=10kW, TAA(0)=26deg

Fig. 7.35 Average change in room temperature deviation ΔE_{SAA} for RegA balancing signal tracking characteristics (communication and processing delay total $\tau_T = 2$ min) suppression change but also the room temperature change response, the effect of the delay characteristics of FastADR control on room temperature management is also quantitatively evaluated.

The price became available. In this simple feed-forward control, the temperature rise was not fed back, and the response of power and room temperature deviation to the power suppression value $L(t)$ was confirmed. In the future, closed loop control of power consumption will be performed while managing temperature comfort with an optimal regulator.

At present, it is considered difficult to perform feedback control in increments of minutes during FastADR implementation time, that is, for a single rotation air-conditioner, even if it is long enough. However, if 5G next-generation mobile communications, etc. are introduced, ultra-high-speed and ultra-low-cost communications infrastructure can be used, and power can be obtained from resource aggregators on the cloud that can implement complex control, or in tens of seconds. A system that can optimally trade off optimal power suppression and room temperature comfort while feeding back the suppression response and room temperature fluctuation response is not a dream.

7.5 Building Multi-type Air-Conditioning FastADR and Room Temperature Preservation

7.5.1 Building Multi-type Air-Conditioning Power/Room Temperature State Space Model

In the virtual power plant simulation performance evaluation for frequency control of the entire power system described above, it was assumed that individual building multi-type air-conditioners gave top priority to power suppression by FastADR command. In that case, power suppression control was not considered in consideration of the fact that the room temperature fluctuated as a result of FastADR power suppression and comfort was impaired.

However, since the air-conditioner is a facility for maintaining room temperature comfort, it is necessary to consider the influence of room temperature fluctuations. Unlike home air-conditioning, office buildings have heat capacity, and in a few minutes, it is highly possible that a rise in room temperature will not compromise comfort. The state space model of modern control theory can be considered as a precise multi-variable mathematical model that expresses this relationship quantitatively.

If the state space model is obtained, the feedback control method can be studied while balancing the state variables in a trade-off relationship, unlike the classical control theory, using the modern control theory.

As mentioned earlier, it is very difficult to identify the physical model of power consumption and building heat load of the building multi-unit air-conditioner. Therefore, we introduce a method to identify the system as a state space representation from the time series measurement data described below via the autoregressive linear model. Since this method is linear, it does not have non-linear expression ability like a neural network, but it gives priority to the merit obtained as state space expression applicable to modern control theory.

This section deals with the response for several minutes of air-conditioning built-in control to the aggregator power suppression command. Therefore, instead of variables related to static air-conditioning heat balance, only power consumption and room temperature representative point changes as responses in the last few minutes and one minute later are used as modeling variables. This AR model has the following structure obtained in previous studies [6, 7, 18].

$$\begin{bmatrix} P(t) \\ T_{SA}(t) \end{bmatrix} = \sum_{l=1}^{L} \begin{bmatrix} a_{11}(l) & a_{12}(l) & a_{13}(l) \\ a_{21}(l) & a_{22}(l) & a_{23}(l) \end{bmatrix} \begin{bmatrix} P(t-l) \\ T_{SA}(t-l) \\ L(t-l) \end{bmatrix} \quad (7.77)$$

where t [min] is the discrete time in 1 min units, $P(t)$ [kW] is the total air-conditioning power for each building, $T_{SA}(t)$[deg] is the average value of each building for room temperature and set temperature deviation, $L(t)$[kW] is the air conditioning power suppression command from the aggregator to each building and corresponds to $L_{\_1}$, $L_{\_2}, \ldots, L_{\_N}$ in Fig. 7.31. $a_{11}(l), \ldots, a_{23}(l)$ are AR coefficient matrix elements, where l is a time series step integer and its maximum value L is the AR model order.

In order to obtain a discrete state space representation from this AR model Eq. 7.7 is divided into a state variable part and an operation variable part as Eq. 7.78.

$$\begin{aligned} \begin{bmatrix} P(t) \\ T_{SA}(t) \end{bmatrix} &= \sum_{l=1}^{L} \boldsymbol{A}_l \begin{bmatrix} P(t-l) \\ T_{SA}(t-l) \end{bmatrix} + \sum_{l=1}^{L} \boldsymbol{B}_l L(t-l) \\ &= \sum_{l=1}^{L} \begin{bmatrix} a_{11}(l) & a_{12}(l) \\ a_{21}(l) & a_{22}(l) \end{bmatrix} \begin{bmatrix} P(t-l) \\ T_{SA}(t-l) \end{bmatrix} + \sum_{l=1}^{L} \begin{bmatrix} a_{13}(l) \\ a_{23}(l) \end{bmatrix} L(t-l) \end{aligned} \quad (7.78)$$

In Eq. 7.78, $\boldsymbol{A}_l$ is the state variable coefficient matrix, $\boldsymbol{B}_l$ is the manipulated variable coefficient matrix (vector). The model construction, so far, is based on the variable selection and the standard time series analysis method [19] according to the previous research [6, 7, 18].

Next, the shift state variable vector is defined to obtain the state space representation from the AR model, which is newly proposed in this study. That is, the state variable vector is defined as $\boldsymbol{x}(t) \triangleq [P(t), T_{SA}(t)]^{\mathrm{T}}$, and the operation variable (scalar) is defined as $u(t) \triangleq P_L(t)$. Here, to convert Eq. 7.78 into a state space representation, we introduce a shift state variable vector $\boldsymbol{x}_k(t)$ shifted by k steps as Eq. 7.79.

$$\boldsymbol{x}_k(t) \triangleq \sum_{l=k+1}^{L} \{\boldsymbol{A}_{l-1}\boldsymbol{x}_k(t+k-l) + \boldsymbol{B}_{l-1}\boldsymbol{u}(t+k-l)\} \tag{7.79}$$

In this AR model identification, since the AR model order is the maximum likelihood order $L = 3$ according to the AIC criterion [19], a shift state vector of $k = 0$, 1, 2 is obtained as Eq. 7.80.

$$\boldsymbol{x}_0(t) = \boldsymbol{A}_0\boldsymbol{x}_0(t-1) + \boldsymbol{x}_1(t-1) + \boldsymbol{B}_0\boldsymbol{u}(t-1) \tag{7.80}$$

$$\boldsymbol{x}_1(t) = \boldsymbol{A}_1\boldsymbol{x}_0(t-1) + \boldsymbol{x}_2(t-1) + \boldsymbol{B}_1\boldsymbol{u}(t-1) \tag{7.81}$$

$$\boldsymbol{x}_2(t) = \boldsymbol{A}_2\boldsymbol{x}_0(t-1) + \boldsymbol{B}_2\boldsymbol{u}(t-1) \tag{7.82}$$

Therefore, by defining the integrated state variable $z(t) \triangleq [\boldsymbol{x}_0(t), \boldsymbol{x}_1(t), \boldsymbol{x}_2(t)]^{\mathrm{T}}$, the Eq. 7.83 is obtained.

$$\begin{bmatrix} \boldsymbol{x}_0(t) \\ \boldsymbol{x}_1(t) \\ \boldsymbol{x}_2(t) \end{bmatrix} = \begin{bmatrix} \boldsymbol{A}_0 & \boldsymbol{I} & 0 \\ \boldsymbol{A}_1 & 0 & \boldsymbol{I} \\ \boldsymbol{A}_2 & 0 & 0 \end{bmatrix} \begin{bmatrix} \boldsymbol{x}_0(t-1) \\ \boldsymbol{x}_1(t-1) \\ \boldsymbol{x}_2(t-1) \end{bmatrix} + \begin{bmatrix} \boldsymbol{B}_0 \\ \boldsymbol{B}_1 \\ \boldsymbol{B}_2 \end{bmatrix} u(t-1) \tag{7.83}$$

Rewriting *z(t)* with Eq. 7.83 as a state equation, the discrete state space representation is obtained as Eq. 7.84.

$$z(t) = \boldsymbol{\Phi} z(t-1) + \boldsymbol{\Gamma} u(t-1) \tag{7.84}$$

$$y(t) = \boldsymbol{H} z(t) \tag{7.85}$$

Here, $\boldsymbol{\Phi}$ is the state coefficient, $\boldsymbol{\Gamma}$ is the input coefficient, H is the output coefficient, and *y(t)* is the output variable. $\boldsymbol{\Phi}$ and $\boldsymbol{\Gamma}$ are obtained from the AR model coefficient matrix elements as Eq. 7.86.

$$\boldsymbol{\Phi} = \begin{bmatrix} \boldsymbol{A}_0 & \boldsymbol{I} & 0 \\ \boldsymbol{A}_1 & 0 & \boldsymbol{I} \\ \boldsymbol{A}_2 & 0 & 0 \end{bmatrix} = \begin{bmatrix} a_{11}(1) & a_{12}(1) & 1 & 0 & 0 & 0 \\ a_{21}(1) & a_{22}(1) & 0 & 1 & 0 & 0 \\ a_{11}(2) & a_{12}(2) & 0 & 0 & 1 & 0 \\ a_{21}(2) & a_{22}(2) & 0 & 0 & 0 & 1 \\ a_{11}(3) & a_{12}(3) & 0 & 0 & 0 & 0 \\ a_{21}(3) & a_{22}(3) & 0 & 0 & 0 & 0 \end{bmatrix}$$

$$\boldsymbol{\Gamma} = [\boldsymbol{B}_0, \boldsymbol{B}_1, \boldsymbol{B}_2]^{\mathrm{T}} \tag{7.86}$$

$$= [a_{13}(1), a_{23}(1), a_{13}(2), a_{23}(2), a_{13}(3), a_{23}(3)]^{\mathrm{T}} \tag{7.87}$$

Thus, after constructing the discrete time-series data to the AR model as Eq. 7.78 by the conventional method, the discrete state space model representation in Eq. 7.86 in modern control theory was obtained by the shift state variable vector proposed in this paper.

7.5.2 Building Multi-type Air-Conditioning FastADR Optimum Regulator

In the previous section, the relationship between FastADR power suppression and room temperature fluctuation was obtained as a state space representation model using state space representation. By obtaining a state space representation model, we were able to simulate the interaction between power suppression and room temperature fluctuation. As a result, we evaluated the simulation of feedforward control that gives a proper power suppression operation amount that is constant during the FastADR duration while maintaining the room temperature within a fixed tolerance (e.g., ± 0.5 °C).

However, in reality, it is difficult to determine in advance a certain power suppression value that can maintain this room temperature comfort. This is because, of course, how far room temperature fluctuations are accepted by restraining power at the request of FastADR changes in relation to FastADR implementation reward incentives.

Therefore, we introduce a method of feedback control from moment to moment during the FastADR duration so as to optimize the evaluation function consisting of the power suppression value given the reward incentive as a parameter and the room temperature fluctuation value.

There are various types of such methods, but in this document we have obtained a state space representation consisting of the power suppression command, actual power, and room temperature deviation, so FastADR feedback optimal control with the most suitable modern control theory regulator is an example. It states as.

The state space representation model described in the previous section of this document is shown again in Eq. 7.88.

$$\begin{bmatrix} P(t) \\ T_{SA}(t) \end{bmatrix} = \sum_{l=1}^{L} \begin{bmatrix} a_{11}(l) \; a_{12}(l) \; a_{13}(l) \\ a_{21}(l) \; a_{22}(l) \; a_{23}(l) \end{bmatrix} \begin{bmatrix} P(t-l) \\ T_{SA}(t-l) \\ L(t-l) \end{bmatrix} + \boldsymbol{E}_{WN}(t) \tag{7.88}$$

t is the separation time, $P(t)$ is the air-conditioning power of the entire building, $T_{SA}(t)$ is the average temperature deviation of the entire building, $a_{ij}(l)$ is the AR model coefficient, l is the number of delay steps, L is the AR model order (this time $L = 3$), $\boldsymbol{E}_{WN}(t)$ is white noise (hereinafter the noise term is omitted).

Since the transition from the AR model to the state space representation model was done in the previous chapter, it is omitted.

$$\begin{bmatrix} \boldsymbol{x}_0(t) \\ \boldsymbol{x}_1(t) \\ \boldsymbol{x}_2(t) \end{bmatrix} = \begin{bmatrix} \boldsymbol{A}_0 & \boldsymbol{I} & 0 \\ \boldsymbol{A}_1 & 0 & \boldsymbol{I} \\ \boldsymbol{A}_2 & 0 & 0 \end{bmatrix} \begin{bmatrix} \boldsymbol{x}_0(t-1) \\ \boldsymbol{x}_1(t-1) \\ \boldsymbol{x}_2(t-1) \end{bmatrix} + \begin{bmatrix} \boldsymbol{B}_0 \\ \boldsymbol{B}_1 \\ \boldsymbol{B}_2 \end{bmatrix} u(t-1) \tag{7.89}$$

$$z(t) \triangleq \begin{bmatrix} \boldsymbol{x}_0(t) \\ \boldsymbol{x}_1(t) \\ \boldsymbol{x}_2(t) \end{bmatrix} \tag{7.90}$$

Here, the state variable vectors can be summarized as discrete state space model in Eq. 7.91.

$$z(t) = \boldsymbol{\Phi} z(t-1) + \boldsymbol{\Gamma} u(t-1) \tag{7.91}$$

$$y(t) = \boldsymbol{H} z(t) \tag{7.92}$$

Here, Φ is the state coefficient, Γ is the input coefficient, H is the output coefficient, and $y(t)$ is the output variable. Φ and Γ are obtained from the AR model coefficient matrix elements as Eqs. 7.93 and 7.94.

$$\boldsymbol{\Phi} = \begin{bmatrix} \boldsymbol{A}_0 & \boldsymbol{I} & 0 \\ \boldsymbol{A}_1 & 0 & \boldsymbol{I} \\ \boldsymbol{A}_2 & 0 & 0 \end{bmatrix} = \begin{bmatrix} a_{11}(1) & a_{12}(1) & 1 & 0 & 0 & 0 \\ a_{21}(1) & a_{22}(1) & 0 & 1 & 0 & 0 \\ a_{11}(2) & a_{12}(2) & 0 & 0 & 1 & 0 \\ a_{21}(2) & a_{22}(2) & 0 & 0 & 0 & 1 \\ a_{11}(3) & a_{12}(3) & 0 & 0 & 0 & 0 \\ a_{21}(3) & a_{22}(3) & 0 & 0 & 0 & 0 \end{bmatrix} \tag{7.93}$$

$$\boldsymbol{\Gamma} = \begin{bmatrix} \boldsymbol{B}_0 \\ \boldsymbol{B}_1 \\ \boldsymbol{B}_2 \end{bmatrix} = \begin{bmatrix} a_{13}(1) \\ a_{23}(1) \\ a_{13}(2) \\ a_{23}(2) \\ a_{13}(3) \\ a_{23}(3) \end{bmatrix} \tag{7.94}$$

Since a discrete state space model of power $P\ (t)$, average room temperature deviation $T_{SA}(t)$, and power limit manipulated variable $L\ (t)$ was obtained, evaluation of $P\ (t)$, $T_{SA}(t)$, and $L\ (t)$ An optimal controller as a function can be designed.

The optimal regulator evaluation function J_{TE} for the discrete time interval ($t = 1, \ldots, T_E$) is defined as follows using $Q\ (t)$ and $R\ (t)$ as evaluation weighting factors.

$$J_{TE} = \sum_{t=1}^{T_E} \left(z(t)^{\boldsymbol{T}} \boldsymbol{Q}(t) z(t) + u(t-1) R(t) u(t-1) \right) \tag{7.95}$$

Here, $z(t)^T \boldsymbol{Q}(t) z(t)$ is the state variable deviation penalty, and $u(t-1)R(t)u(t-1)$ is the excessive manipulated penalty.

Assume that the evaluation weight coefficient *Q (t)* and only the latest state variable need to be evaluated. The evaluation function has three weighting factors q_{11}, q_{22}, and *R*, and the evaluation function J_{TE} is defined as Eq. 7.96.

$$\boldsymbol{Q}(t) = \begin{bmatrix} q_{11} & 0 & \cdots & 0 \\ 0 & q_{22} & \cdots & 0 \\ \vdots & \vdots & \ddots & \vdots \\ 0 & 0 & \cdots & 0 \end{bmatrix} \tag{7.96}$$

$$\boldsymbol{J_{TE}} = \sum_{t=1}^{T_E} \left(\boldsymbol{q}_{11} \boldsymbol{P}(t)^2 + \boldsymbol{q}_{22} \boldsymbol{E}_{SA}(t)^2 + \boldsymbol{RL}(t-1)^2 \right) \tag{7.97}$$

Approximating the steady state in the evaluation interval gives the solution matrix *P* of the discrete-time Riccati equation, as Eq. 7.98.

$$\boldsymbol{P} = \boldsymbol{\Phi}^T \boldsymbol{P} \boldsymbol{\Phi} - \boldsymbol{\Phi}^T \left(\left(\boldsymbol{P\Gamma} \left(R + \boldsymbol{\Gamma}^T \boldsymbol{P\Gamma} \right)^{-1} \boldsymbol{\Gamma}^T \boldsymbol{P} \right) \right) \boldsymbol{\Phi} + \boldsymbol{Q} \tag{7.98}$$

The optimal control state feedback gain matrix *G* is obtained from the optimal solution matrix *P*, and the optimal manipulated variable *u (t)* can be output as Eqs. 7.99 and 7.100.

$$\boldsymbol{G} = -\left(\boldsymbol{R} + \boldsymbol{\Gamma}^T \boldsymbol{P\Gamma} \right)^{-1} \boldsymbol{\Gamma}^T \boldsymbol{P\Phi} \tag{7.99}$$

$$u(t) = \boldsymbol{G}z(t) = [g_1, \ldots, g_6] \begin{bmatrix} \boldsymbol{x}_0(t) \\ \boldsymbol{x}_1(t) \\ \boldsymbol{x}_2(t) \end{bmatrix} = [g_1, \ldots, g_6] \begin{bmatrix} P(t) \\ T_{SA}(t) \\ \boldsymbol{x}_1(t) \\ \boldsymbol{x}_2(t) \end{bmatrix} \tag{7.100}$$

Figure 7.36 shows the result of trial of FastADR control simulation for $T_E =$ 15 min with the optimal regulator using the following evaluation function J_{TE} based on the measured data of the office building.

Unlike the feedforward fixed control described in the previous section, the FastADR power suppression command value is not directly applied to the air-conditioner as the manipulated variable. As can be seen from the simulation result graph in Fig. 7.36, the evaluation function is fed back while the room temperature deviation is fed back. It can be seen that the amount of operation is changed from moment to moment so that is minimized. Immediately after the start of FastADR, a large command value is output so as to approach the power limit command quickly. At this point in time, the room temperature deviation has not deteriorated so much, so that a large command value is eventually fed back so that the evaluation function can be optimized throughout the FastADR.

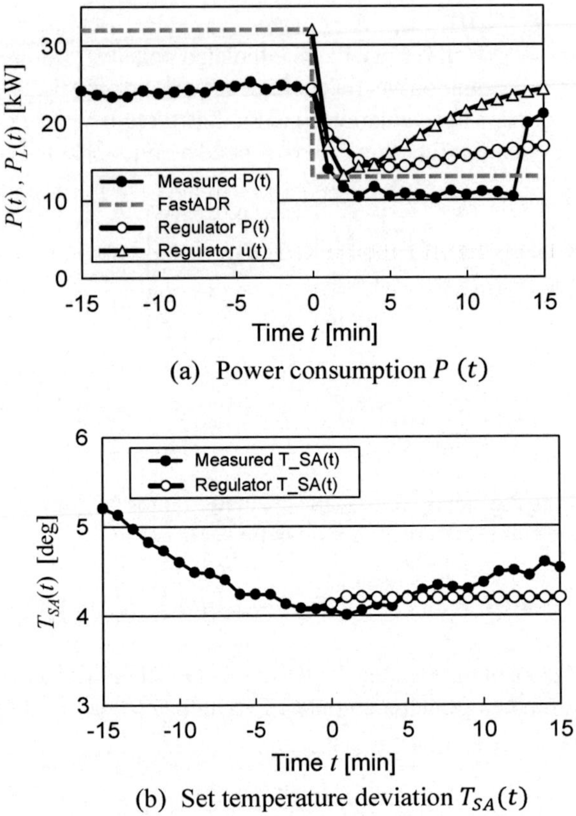

Fig. 7.36 Comparison of measured data for prediction of building air-conditioning FastADR optimal regulator

When the power decreases rapidly and approaches the FastADR power suppression command value, it can be seen that the amount of operation is gradually returned to relax the power suppression. Originally, with conventional feedforward fixed control, the room temperature deviation gradually starts to worsen. However, since the optimum regulator gradually returns the operation amount and relaxes the power suppression, the room temperature deviation can maintain the room temperature at the start of FastADR. In this example, since the penalty parameter of the evaluation function is focused on room temperature, even if the power suppression performance is slightly different from the FastADR power suppression command value, the evaluation function is set to be the optimum value when the room temperature is maintained. If the highest priority is to bring the response power close to the FastADR power suppression command value even if the room temperature deteriorates, the penalty parameter of the evaluation function should be allocated in that way.

In this way, by adjusting the trade-off between the incentive economy for the accuracy of response to the FastADR response and the importance of maintaining the room temperature in the air-conditioned space, a control that trades off power saving priority or room temperature maintenance priority to an appropriate ratio. The method can be realized.

7.6 Virtual Power Plant Communication and Traffic Evaluation

7.6.1 FastADR Web Service Communication Analysis

One point to realize a large-scale virtual power plant is that the supply and demand can be adjusted by aggregating hundreds and thousands of customer facilities distributed over a wide area. In the economic operation control (EDC) control from the central power supply command station, it is necessary to have a response of several minutes as described in the previous chapter in order to control it equivalent to the existing thermal power plant is there. In order to realize this, in terms of communication performance, the issue is whether it is possible to aggregate information exchange with a large amount of distributed power resources on the order of about 1 min.

In the entire virtual power plant system, the OpenADR communication part between the power system operator of the power company and the top-level aggregation coordinator of the virtual power plant is limited, so the performance can be secured by adopting a dedicated line.

For example, there is a report that FastADR of OpenADR that collects 100 customers can be completed in about 2 s on average [16]. In addition, since the OpenADR 2.0 standard describes 4 s as an example of the FastADR time [13], this upstream part is secured by high-speed communication using a dedicated line, which is a major problem with communication delay. It will not be.

Therefore, the problem of the entire virtual power station communication system is the aggregation part of the downstream resource aggregator and consumers. This part needs to exchange information with hundreds or thousands of customers using low-cost communication infrastructure. In the near future, it would be reasonable to assume that the communications infrastructure will be Internet-based wide area web service communications.

When using the Internet, the HTTPs protocol is used because of the problems of modifying the firewalls of each customer, passing through without discussing the details, and the interconnectivity between the resource aggregator and each energy resource controller. Web service communication base will be one of the realizations [20, 21].

In this section, we examine the data transmission characteristics of the IEEE1888 standard [14] as a communication framework for aggregation Web service communication. This communication framework standard is an IEEE/ISO standardized framework developed specifically for Internet equipment monitoring, as introduced in Chap. 3. Since this standard is premised on Web service communication using

HTTP (S), it seems that a system with excellent interoperability can be constructed for various customers [21, 22]. However, it is important to note that because HTTP (S) over TCP/IP is used, Internet data transmission time may take several seconds in some cases, and the data transmission delay is probabilistically distributed. Yes [15].

As a target example of delay time for aggregation web service communication, typical providers in the industry assume that web service communication is completed with each customer within one minute [24]. In addition, the variation in response time with the building equipment side was assumed to be 30 s, which is three times the maximum value, based on an actual measurement with a standard deviation of 9.5 s in the previous study [23]. In other words, considering the maximum delay of less than 1 min, the maximum variation of 30 s was considered, and in this section, the expected delay value was considered to be within 30 s.

In addition to the data transmission delay, the aggregator must consider the processing capacity of the server in order to deal with many customers. In recent years, computers are inexpensive and can be processed in parallel by increasing the number of servers. However, the purpose of the aggregator is to collect the amount of restraint from each customer up to the target amount of restraint and achieve power restraint. When the customers respond to all of them at once, requesting restraint, The possibility of the problem that the power control becomes oscillating must be avoided. For this reason, there may be appropriate situations in which the amount of restraint from each customer is accumulated and the target amount of restraint is negotiated in sequence with the customer while performing aggregation web service communication.

In this section, in order to examine the scalability of the aggregation web service communication system, each aggregation server requests the customer to request the amount of power reduction by using the web service communication of IEEE1888 and obtains an answer. After that, we assumed a sequential accumulation method that requires the next customer. For speeding up in parallel, the study is based on the premise that it will be added in parallel for each aggregation server.

Figure 7.38 shows the queuing theory applied to the target part (the part circled in Fig. 7.37) of the aggregation system in this section. In other words, we consider that there is a waiting room in the aggregation server inside the aggregator, and treat FastADR power suppression commands in BEMS units as customers in the queue. One service time is from the power suppression request from the aggregation server to the BEMS to the response to the power suppression possible from the BEMS to the aggregation server.

Figure 7.39 shows the queue arrival model for the number of FastADR Messages to the aggregation server. This section assumes demand response as a rare event that occurs once a day or not. The FastADR announcement is sent from the power company server Virtual Top Node (VTN) to the aggregator client Virtual End Node (VEN) via a dedicated line. VEN Client and Allocator allocates the total required amount to the equipment load group and decomposes it into a demand response Allocation Message group. It is assumed that the messages are sequentially input from the VEN Allocator to the aggregation server in bursts in a short time of seconds.

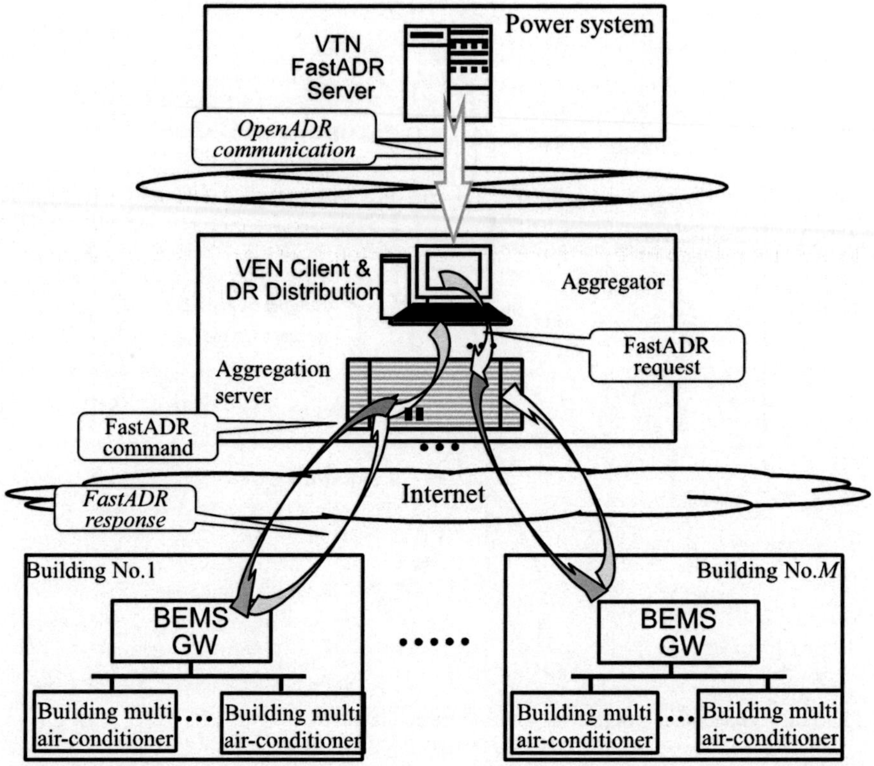

Fig. 7.37 Modeling to analyze traffic of building multi air-conditioning FastADR aggregation communication system in virtual power plant

Therefore, the arrival traffic to the aggregation server is an isolated short burst-like Poisson arrival probability distribution model.

In this section, the M/M/1 type queuing model is used. Various queuing models with more complex arrival probability distributions, service processing probability distributions, and queuing structures have been proposed, but the probability distribution of queue lengths can be expressed analytically with a single formula, including transient states. An analysis example using the M/M/1 type will be described.

The M/M/1 queuing model is a model in which the arrival probability distribution is Poisson distribution, the service time probability distribution is exponential distribution, and the waiting room length is infinite. Since arrival and service are Markovian, both arrival and service are represented by the symbol M. The last symbol 1 means that only one service is processed at a time.

Naturally, the actual FastADR aggregation server communication process is a more complex process. There is a more realistic queuing model, but the only model that can solve transients analytically is M/M/1. This section focuses on examining the processing capacity when the FastADR power suppression command arrives from

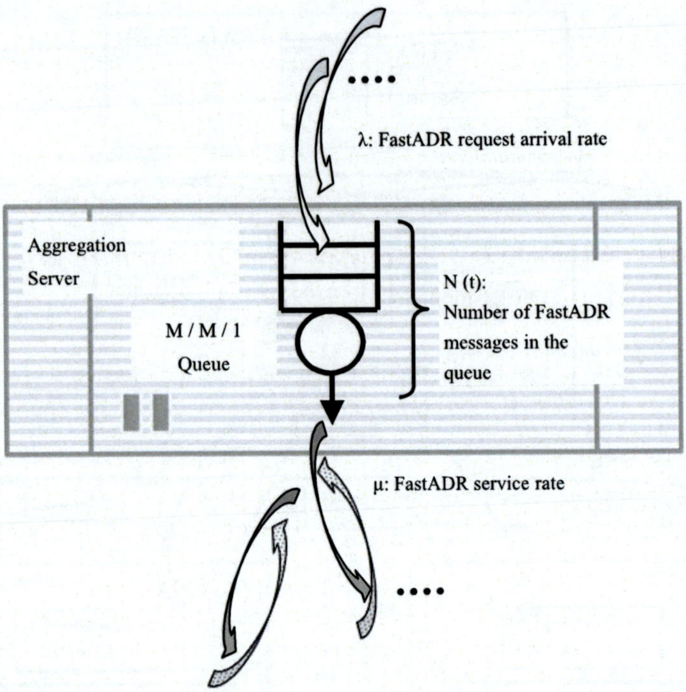

Fig. 7.38 Queue model for internal web service communication in the send buffer in the resource aggregator server

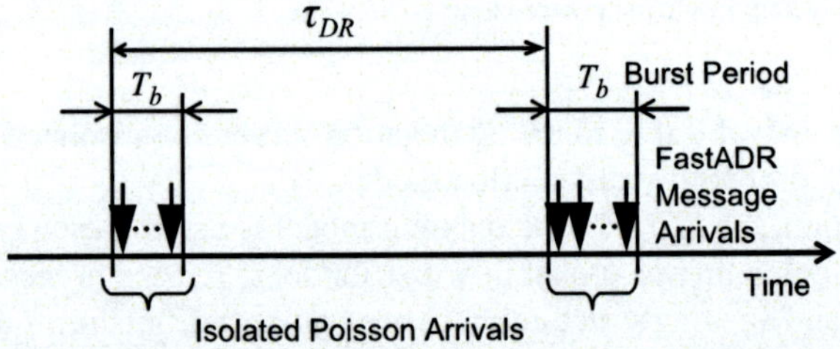

Fig. 7.39 Modeling the arrival time interval of the FastADR power suppression command that arrives in burst form at the aggregation server

the upstream in a bursty manner. Therefore, the M/M/1 queuing model is adopted with the highest priority being able to describe transient phenomena.

The M/M/1 queue can describe the transient solution of its state equation as an analytical expression. $P_{j,r}$ $(t) = \mathrm{P}\{N(t) = \mathrm{r}$ I Probability that the number $N(t)$ of aggregation Web service communication messages in the queue at time t is $N(0) = j$, $N(t) = r$ The equation of state for $N(0) = j\}$ is as Eq. 7.101, from [25, 26].

$$\frac{d}{dt}P_{j,r}(t) = \begin{cases} -\lambda P_{j,0}(t) + \mu P_{j,1}(t) \quad (r = 0) \\ \lambda P_{j,r-1}(t) - (\lambda + \mu)P_{j,r}(t) + \mu P_{j,r+1}(t) \quad (r \geq 1) \end{cases} \tag{7.101}$$

The transient solution of the above state equation (Eq. 7.101), that is, the probability $P_{j,r}(t)$ that the number $N(t)$ of messages in the queue is j in the initial state and r at time t can be calculated by [25, 26] in Eq. 7.102.

$$P_{j,r}(t) = \exp[-(1+\rho)\mu t] \cdot \rho^{\frac{r-j}{2}} \cdot \Bigg\{ J_{r-j} + \rho^{-\frac{1}{2}} \cdot J_{r+j+1} + (1-\rho)\sum_{k=1}^{\infty}\Big[\rho^{-\frac{k+1}{2}} \cdot J_{r+j+k+1}\Big]\Bigg\} \tag{7.102}$$

$$J_n = I_n\big(2\mu t\sqrt{\rho}\big) = I_n\Big(2t\sqrt{\mu\lambda}\Big) \tag{7.103}$$

where ρ is the product of arrival rate λ and service rate μ, and $I_n\big(2t\sqrt{\mu\lambda}\big)$ is the *nth*-order Bessel function.

Therefore, the expected value E[$N(t)$] of the number of messages in the queue at time t is given by multiplying the probability of r messages at time t and summing them up as Eq. 7.104.

$$\mathrm{E}[N(t)] = \sum_{r=0}^{\infty} P_{0,r}(t) \cdot r \tag{7.104}$$

Let M be the number of destination buildings in the overall system, that is, the total number of communication devices to be sent, such as the destination energy resource controller (ERC) or building energy management system (BEMS).

Figure 7.40 conceptually shows the transmission processing status of the FastADR power suppression command in the transmission buffer in the aggregation server.

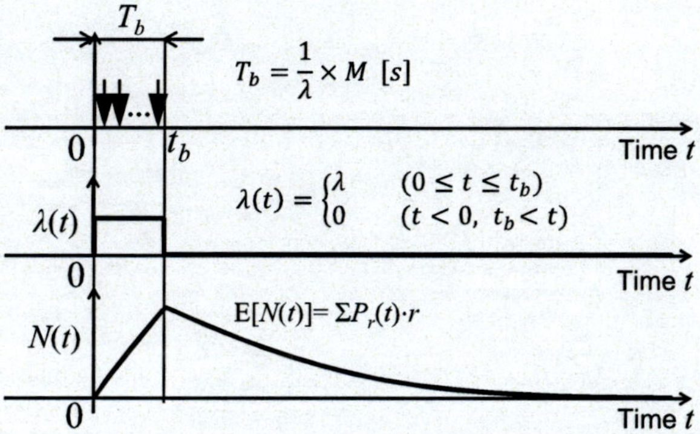

Fig. 7.40 Transient change in the number N (t) of waiting transmission commands in the FastADR command group that arrives in bursts and randomly in the M/M/1 queuing model

The time $t = 0$ is the Web service communication request start time, and the number of customers in the system at that time is $N(0) = 0$. FastADR power suppression commands are sent less frequently, and FastADR aggregation is performed with sufficient time intervals. In other words, it is assumed that communication packets arrive in a queue in bursts.

The FastADR Message arrival burst time Tb in this analysis is the time from $t = 0$ to the time tb when all FastADR messages arrive from the VTN. Also, let the arrival rate of this burst time Tb be λ, and the arrival rate at other times be 0.

Theoretic analysis requires parameter values for the arrival rate λ at the burst arrival time t_b and the service rate μ of the aggregation web service communication. λ and μ are measured using a test facility described in the next chapter with a dummy Internet environment and actual Web service communication software. Since this analysis model assumes the steady state in steps as shown in Fig. 7.40, the steady state value was determined by the average of 1000 measurements. It is assumed that λ is experimentally determined in a laboratory LAN environment and μ is experimentally determined in a dummy internet environment described later.

7.6.2 FastADR Web Service Communication Simulation

It is difficult to carry out FastADR aggregation web service communication experiments using large-scale facilities distributed in a wide area. Therefore, this section describes a method for evaluating data transmission communication traffic for actual aggregation communication without using a real machine. Specifically, the real-time simulation (emulation) experimental facility shown in Fig. 7.41 is shown as

Fig. 7.41 Example of real-time simulation (emulation) experimental facility for building multi air conditioning FastADR aggregation communication

an example. This experimental facility is an experimental facility that combines the power grid supply and demand introduced in the previous section and a building multi-type air-conditioning group virtual power plant by executing simulation in real time.

In this simulation experiment facility, FastADR aggregation Web service communication was actually implemented and constructed by a total of about 100 computers. This facility faithfully implements the actual aggregation web service communication, and can reproduce the actual delay time occurrence probability and conduct experiments.

In this facility, the typical transmission conditions of the long-distance Internet were reproduced in order to simulate the stochastic delay of the Internet and the probability of IP packet transmission errors. Specifically, the Dummynet emulator [27], which is well-known as an IP packet traffic emulator for reproducing a wide area network environment, is used.

In the experiments in this section, the packet loss probability of Dummynet is generated by the Gilbert-Elliot loss model [28, 29]. The Gilbert-Elliot loss model can simulate behavior similar to packet loss in an actual Internet environment where packet loss occurs in bursts. The packet loss occurrence rate PL as a long-term average is 2% of the typical Internet environment [20, 23].

In the experiment, we consider two cases, the case of low delay and the case of large delay between the aggregation server and BEMS, and conduct the experiment 100 times each, and compare the average and the expected value of theoretical analysis.

Figure 7.42 shows the FastADR aggregation web service communication sequence in this experimental facility. As described above, the aggregation web

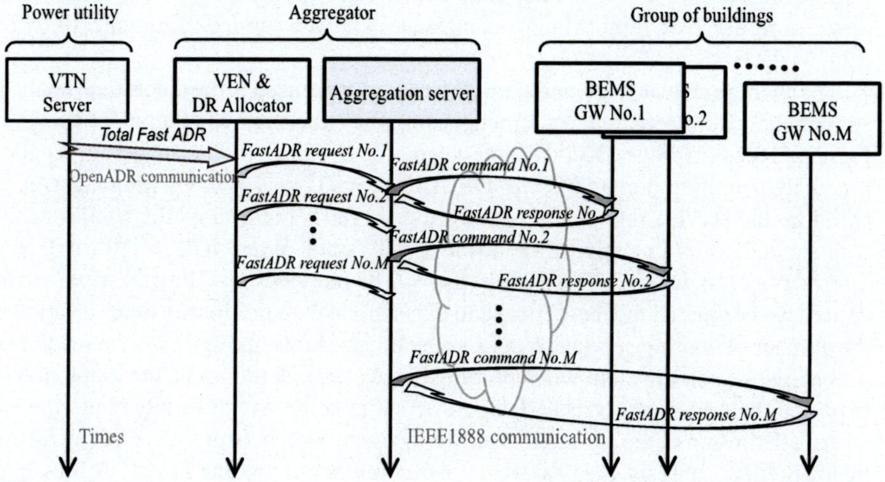

Fig. 7.42 Conceptual diagram of overall sequence of FastADR aggregation communication system for building multi air conditioner

service communication between the aggregation server and each BEMS is implemented using the IEEE1888 standard. Specifically, it uses IEEE1888 TRAP Web Service, which enables "server push" to notify demand response requests from the server side. This is a Web service communication suitable for wide-area remote control, in which the BEMS on the client side first registers the desired conditions in the aggregation server on the server side and can notify the client from the server when the conditions are satisfied.

Conversely, to report power suppression results from the building side, it is implemented using the normal IEEE1888 W RITE Web Service that writes from the client to the server. In this section, the OpenADR communication part on the VTN Server side, which is not covered in this section, is configured in such a way that VEN Client & DR Allocator, which is a higher-level function of the resource aggregator, also takes on that function.

Experimenting with the FastADR aggregation Web service communication system using the wide-area monitoring and control Web service communication method of the standard that can be used at this stage so that it can be extended to FastADR aggregation with a large number of customers in this way It is built indoors.

In this section, FastADR wide area Web service communication assumes the delay condition that other communication coexists in general commercial Internet. Considering future communication technology development, the present Internet field test experimental value was considered as the worst value in the future. As experimental conditions, typical and long-distance Internet environment conditions are $RTT = 10$ [ms] and 100 [ms] for the IP packet round-trip delay time, and $P_L = 2\%$ for the IP packet loss rate. These RTT and P_L are the minimum and maximum typical values in Japan and overseas based on Internet RTT measurement data in Nagoya-Tokyo and Nagoya-Bangkok, which were measured in field tests in our previous research on remote building management web service communication [20] (Fig. 7.43).

The average arrival rate λ and average service rate μ used in the theoretical analysis are obtained from actual measurements using the experimental equipment shown in Fig. 7.41. The average FastADR message arrival rate arriving at the aggregation server PC from the higher VEN & DR Allocator PC was $\lambda = 9.4$ [messages/s]. In addition, the service rate of the aggregation server is the average of 1000 times of web service communication time in the steady state. When $RTT = 10$ [ms], $\mu = 1.46$ [services/s], when $RTT = 100$[ms], $\mu = 0.82$ [services/s]. Using these parameter values, we compared the theoretical and experimental values for the time variation of the number of waiting services N(t). As a result, as shown in Fig. 7.42, the results are probabilistic each time, but when plotting the average of 100 trials, the graph of N(t) expected theoretical analysis and experimental average Are generally consistent.

As a definition of the parameters used for comparison between the experiment and the theoretical analysis, the peak value of the number of waiting FastADR Messages N(t) in the aggregation server shown in Fig. 7.44 is defined as Npeak. Also, the time until N(t) decreases to 10% or less after the arrival of FastADR Message is defined as the aggregation web service communication completion time T_{AG}. Strictly speaking, the time should be 0%, but in theoretical analysis, N(t) will not be completely zero

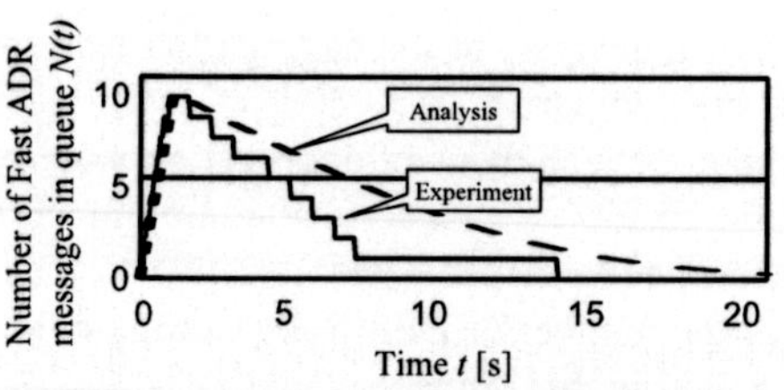

(a) Comparison of the experimental value of one time and analytical solution

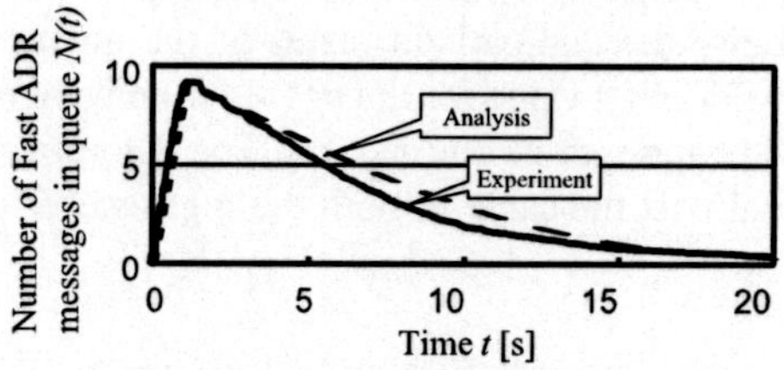

(b) Comparison of 100 experimental averages and analytical solutions

Fig. 7.43 Comparison of average value and queuing analysis value of repeated experiment of FastADR communication system real-time simulator

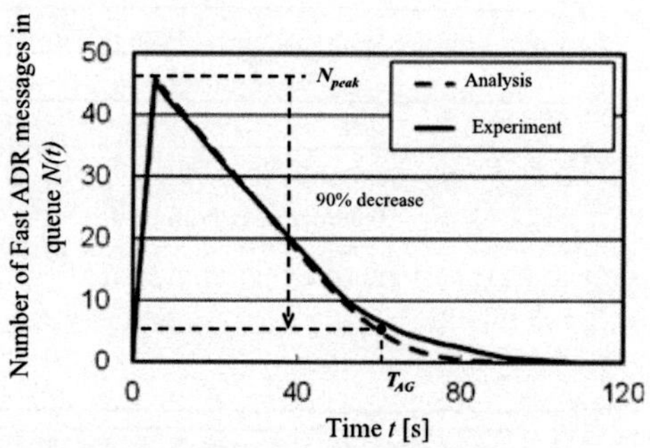

Fig. 7.44 Comparison of queuing analysis solution and simulation experiment for transient phenomenon of FastADR command waiting for transmission in aggregation server

unless infinite time is reached. As a realistic system design, it is sufficiently meaningful that the application layer recovers even if the communication layer timeout processing is based on this 10% expectation value, so this definition is sufficiently meaningful.

The network environment between the aggregation server of the aggregator and the energy resource controller or BEMS includes the two cases described above: Case 1 (High-speed Internet: $RTT = 10$ [ms]) and Case 2 (Low-speed Internet: $RTT = 100$ [ms]) Let's consider about. In each case, the aggregation server increases the number of subordinate buildings per unit, that is, the number of partner BEMS M, and the experiment and theoretical analysis of the aggregation web service communication completion time TAG are compared.

We compare the theoretical analysis of aggregation web service communication described above and the results of web service communication experiment. As the scale of the experimental equipment, it was possible to measure up to 50 energy resource controllers or the number of BEMS M. Compare the theoretical value and the experimental value within this range.

First, we can confirm that the queue peak value Npeak in the aggregation server can be predicted with an error within a few percent under all conditions. Next, Fig. 7.45a and b shows the comparison result of the queue model analysis value and the emulation experiment value for the aggregation web service communication completion time T_{AG} important in this section. The theoretical analysis and experimental results were almost the same in both the high-speed Internet case 1 and the

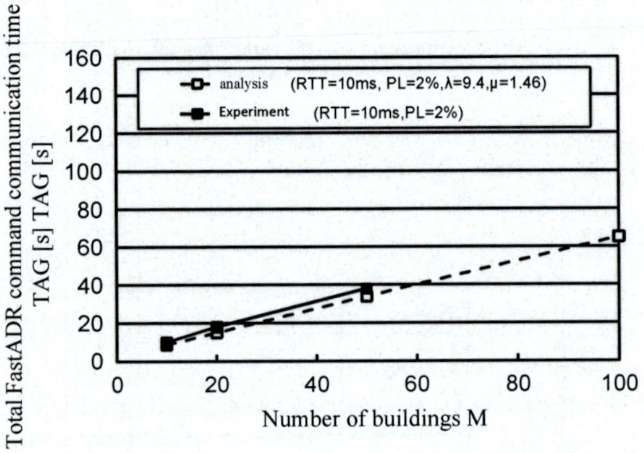

(a) Case1: Fast internet environment, $RTT = 10$ [ms]

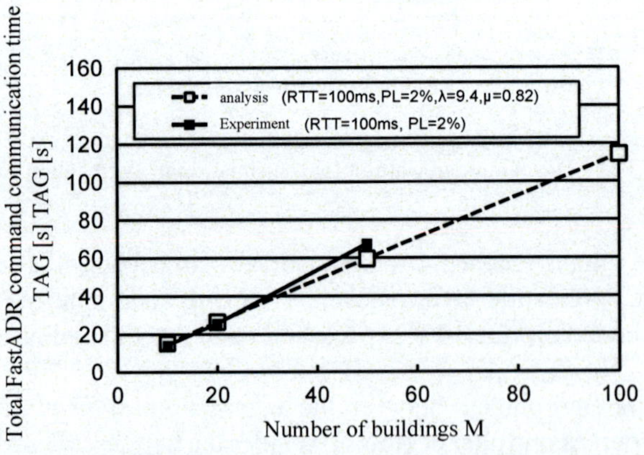

(b) Case2: Slow internet environment, $RTT = 100$ [ms]

Fig. 7.45 Scalability analysis for the total number of destinations related to the complete FastADR command transmission completion time of the aggregation server low-speed Internet case 2. The theoretical analysis in this section can be said to be at a level where useful information can be obtained as an aid to quantitative estimation up to the number of buildings $M = 100$ when considering the scalability of the aggregation Web service communication time.

In this section, we considered a 100 [MW] class FastADR aggregation system equivalent to a thermal power plant as a specific scalability study case. This reduced demand of 100 MW is about 1000 units of air-conditioning power for a typical medium-sized office building, for example, if the air-conditioning power of an office building covers the demand response. Since it is impossible to conduct an experiment in such a large-scale case, let us examine it by the theoretical analysis of the queue developed in this section.

If the aggregation web service communication completion time T_{AG} is predicted by theoretical analysis, even if there is no packet loss rate ($PL = 0$ [%]) for low-speed Internet ($RTT = 10$ [ms]) $T_{AG} = 761$ [s] was also necessary. Since the load on each building demand side starts to suppress power after this Web service communication arrives, the power demand can be reached to the target level in 1 min, which is the requirement of FastADR for the purpose of power demand suppression service that responds within minutes after the announcement. Immediate responsiveness cannot be achieved.

Therefore, improvements such as parallel execution of aggregator aggregation servers are necessary. When examining the parallelization of the FastADR aggregation system, if the number of BEMS in charge is $M = 40$, as shown in Fig. 7.45a, the theoretical analysis yields 27 [s]. This is a scalability estimate that, for example, 1000 buildings can be paralleled into 25 systems of 40 buildings in parallel and the aggregation web service communication completion time of 30 [s] set as the target can be almost satisfied.

However, a configuration in which one aggregation system is installed in every 40 buildings would need to be considered for capital investment. This time, the IEEE1888 communication protocol was assumed by the progressive suppression stacking method, but it is also possible to devise communication protocols such as requesting FastADR all at once and distributing compliance responses with random number delays. In this case, scalability analysis like this section is useful.

There is a delay until the device actually lowers the power and BEMS reports the power time series to the aggregator. In our research [30], we show the result of a simulation experiment [30] modeled based on the actual measured data that was sequentially stopped by demand response request/reply Web service communication in an actual building multi-type air-conditioning system.

Figure 7.46 shows an example of the previous research simulation experiment. In Fig. 7.46, $t0$ to $t1$ is the FastADR aggregation total communication time T_{AG} analyzed in this document, and the subsequent $t1$ to $t2$ time is TBW when the power suppression operation of all 100 air-conditioners is completed. In this example, T_{AG} takes 102 [s] and T_{BW} takes 252 [s]. In this theoretical analysis, only the T_{AG} part is considered, and when it is read in Fig. 7.44b, it corresponds to $T_{AG} = 117$ [s]. This example is an example of a building multi-type air-conditioning system, but the

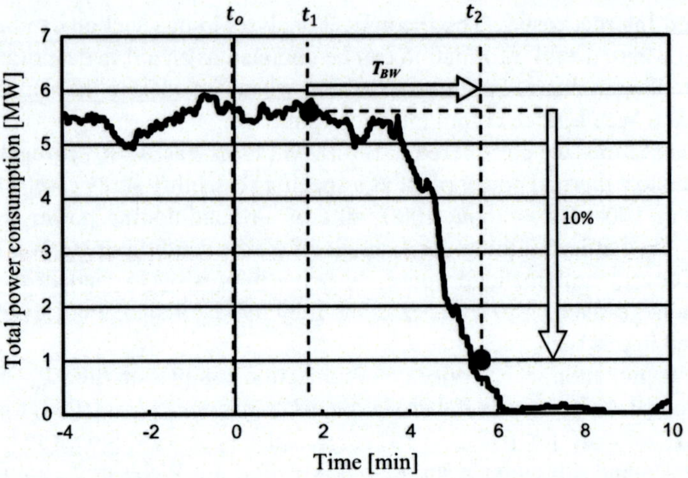

Fig. 7.46 Example of simulation experiment results of FastADR aggregation communication for 100 building multi air conditioners

FastADR total time is $T_{AG} + T_{BW}$, and the communication part T_{AG} covered by this document is about 102/(102 + 252), that is, about 30 [%] Knowledge is obtained.

Demand response request/reply from the aggregator, the command value by Web service communication reaches each customer, and the power reduction process after that varies depending on the actual equipment. Here, the purpose of this section is not included in the theoretical analysis of scalability by a general unified model. The delay in the actual power suppression process is an important issue, but in the future, the power suppression model of a building multi-type air-conditioner, which can be easily standardized as a unified load model, will be combined with the aggregation delay analysis in this section to provide meaningful scalability. It is necessary to analyze it.

References

1. J. Morikawa, C. Ninagawa, Fast real-time pricing optimization control of power consumption and room temperature for building multi-type air-conditioning facilities. IEEJ Trans. Power Energy **136**, 817–823 (2016). ((in Japanese))
2. Y. Aoki, H. Ito, C. Ninagawa, J. Morikawa, T. Inaba, S. Kondo, Complex coordination control with entire power limitation and partial temperature priority of building—Multi-type air-conditioning facilities for real-time pricing. IEEJ Trans. Ind. Appl. **138**(10), 777–786 (2018) (in Japanese)
3. Japan Meteorological Agency. http://www.data.jma.go.jp/obd/stats/etrrn/index.php
4. Technical report of the Inst. of Electrical Engineers of Japan, Standard Analysis Model for Power Supply and Demand/Frequency Simulation, No. 1386 (2016)
5. H. Ito, C. Ninagawa, J. Moriakwa, Fast real-time pricing optimization control with dynamic prediction model for office building facilities, in *IEEJ Technical Meeting on Smart Facilities,*

SMF-16-049 (2016), pp. 79–84 (in Japanese)
6. T. Naramura, J. Morikawa, C. Ninagawa, Dynamic characteristics of side effect of room temperature in FastADR control of multi-type building air-conditioners facilities, in *IEEJ Power and Energy Society Conference* (2015) (in Japanese)
7. T. Naramura, J. Morikawa, C. Ninagawa, Prediction model on room temperature side-effect due to FastADR aggregation for a cluster of building air-conditioning facilities. IEEJ Trans. Power Energy **136**(4), 432–438 (2016) (in Japanese)
8. PJM Manual 12, *PJM Balancing Operations* (Aug, 2016)
9. PJM Web Site Documents, *Normalized Signal Test: RegA*, http://www.pjm.com/markets-and-operations/ancillary-services.aspx (Aug, 2014)
10. Y. Lin, P. Barroah, S. Meyen, T. Middlekoop, Experimental evaluation of frequency regulation from commercial building HVAC systems. IEEE Trans. Smart Grid **6**(2), 776–783 (2015)
11. K. Ma, G. Hu, C. Spanos, Distributed energy consumption control via real-time pricing feedback in smart grid. IEEE Trans. Control Syst. Technol. **22**(5), 1907–1914 (2014)
12. PJM Web Site Documents, *40 Minute Performance Score Template*, http://www.pjm.com/markets-and-operations/ancillary-services.aspx (Aug, 2013)
13. OpenADR Allience, *OpenADR 2.0 Profile Specification B Profile*, Document No. 20120912-1 (2013)
14. IEEE Standard 1888-2011, *IEEE Standard for Ubiquitous Green Community Control Network Protocol* (IEEE, New York, 2011)
15. C. Ninagawa, H. Yoshida, S. Kondo, H. Otake, Data transmission of IEEE1888 communication for wide-area real-time smart grid applications. Int. Renew. Sustain. Energy Conf., IRSEC2013, 1–6 (2013)
16. T. Iwahara, T. Yamada, C. Ninagawa, Smart grid OpenADR communication realized by using IEEE 1888 TRAP mechanism's flexible sever-push capability. IEEJ Technical Meeting on Smart Facilities, SMF-15-14 (2015), pp. 59–63 (in Japanese)
17. C. Ninagawa, T. Iwahara, K. Suzuki, Enhancement of OpenADR communication for flexible FastADR aggregation using TRAP mechanism of IEEE1888 protocol, in *IEEE International Conference on Industrial Technology ICIT2015* (2015), pp. 2450–2454
18. Y. Iwata, K. Saito, C. Ninagawa, J. Morikawa, S. Kondo, Time series combined statistical modeling on dynamic response characteristics of ADR power curtailment for building facilities. IEEJ Indus. Appl. Soc. Conf. v_265–v_266 (2014) (in Japanese)
19. G. Box, G. Jenkins, G. Reinsel, *Time Series Analysis: Forecasting and Control*, 3rd edn (Prentice Hall, 1994), p. 201
20. T. Sato, Y. Kawakita, C. Ninagawa, Analysis for data transmission performance of WEB air-conditioner monitoring system over the internet. TRANS of IEIEJ, **29**(6), 455–460 (2009) (in Japanese)
21. H. Ochiai, M. Ishiyama, T. Momose, N. Fujiwara, K. Ito, H. Inagaki, A. Nakagawa, H. Esaki, FIAP: Facility information access protocol for data-centric building automation systems, in *Proc. IEEE INFOCOM 2011, Workshop on M2MCN-2011* (2011), pp. 229–234
22. H. Ochiai, Power data management on the internet space: green ICT project in Japan, in *IEEE Colombian Communication Conference COLCOM2012* (2012), pp. 1–2
23. T. Yamada, J.Morikawa, C. Ninagawa, Impact of burst packet loss on communication of remote monitoring system for building facilities. TRANS IEIEJ **35**(3), 212–218 (2015) (in Japanese)
24. K. Schisler, T. Sick, K. Brief, The role of demand response in ancillary services markets. IEEE/PES Trans. Distrib. Conf. **2008**, 1–3 (2008)
25. L. Kleinrock, *Queueing Systems. Volume 1: Theory* (Wiley, New York, 1975), pp. 74–77
26. T. Robertazzi, *Computer Networks and Systems: Queueing Theory and Performance Evaluation: Third Edition* (Springer-Verlag, New York, 2000), pp. 68–72
27. L. Rizzo, Dummynet: A simple approach to the evaluation of network protocols. ACM Comput. Commun. Rev. **27**(1), 31–41 (1997)
28. E. Gilbert, Capacity of a burst-noise channel. Bell Syst. Tech. J. **39**, 1253–1265 (1960)

29. E. Elliot, Estimation of error rates for codes on burst-noise channels. Bell Syst. Tech. J. **42**, 1977–1997 (1963)
30. K. Suzuki, C. Ninagawa, Time characteristics of demand response by large-scale aggregation control for building air-conditioning facilities, in *IEEJ Production Facility Control, PFC-14-008* (2014) (in Japanese), pp. 39–44

Chapter 8
Communication Standards in Virtual Power Plant

8.1 OpenADR Communication Standard

8.1.1 Communication Method of OpenADR Standard

Data such as demand response events in OpenADR is exchanged in XML format and messages are defined in XML Schema Definition Language (XSD) [1]. There are two query method PUSH type and PULL type.

The difference between PUSH type and PULL type is the communication protocol corresponding to the subject when the message is transmitted from VTN to VEN. In the case of PULL, VEN is main component, and message notification is periodically requested (polled) to the VTN, and HTTP is used as the communication protocol. In the case of the PUSH type, the VTN is the main component, the message is directly notified to the VEN, and HTTP or XMPP (Extensible Messaging and Presence Protocol) [2] is used as the communication protocol.

HTTP (Hypertext Transfer Protocol) is a communication protocol for exchanging Web pages described in HTML (Hyper Text Markup Language), but not limited to Web pages, also used in various fields for exchanging data. Since HTTP does not include security functions. For security feature, HTTPS (Hypertext Transfer Protocol Secure) combined with TLS is often used.

In addition, it is a communication protocol that can bi-directionally communicate data written in XMPP in real time. Originally developed by adding security functions to the communication protocol used in the instant messaging app called **Jabber**.

TLS (Transport Layer Security) is a partial protocol used to create secure channels on the Internet, and is used in combination with other protocols. The combination of algorithms used for encryption is called a **cipher suite**.

The reason why these inquery methods are used is that there is a situation in which automatic demand response transmits information from VTN to VEN, such as a demand response event. Needless to say, there is no problem in PUSH type even in such situation, but even the PULL type can be transmitted in real time by polling.

C. Ninagawa, *Virtual Power Plant System Integration Technology*, Power Systems,
https://doi.org/10.1007/978-981-16-6148-8_8

Table 8.1 Communication methods in OpenADR

Communication method	Communication messages	Inquiry method	Communication protocol
Simple HTTP PULL	XML defined by XSD	PULL type	HTTP
Simple HTTP PUSH		PUSH type	
XMPP			XMPP

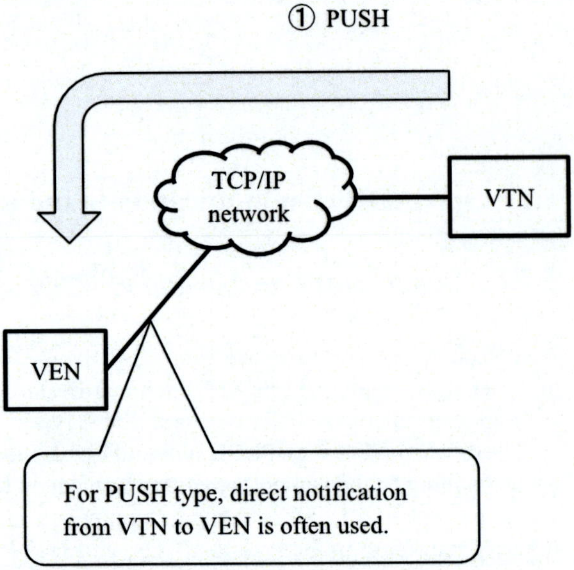

Fig. 8.1 OpenADR PUSH communication sequence

In addition, these inquery methods and communication protocol combinations have names, which are shown in Table 8.1. Figure 8.1 shows an example of a communication sequence for each inquiry method, and Fig. 8.2 shows an example of a communication message for XMPP.

In order to secure these messages, OpenADR requires TLS 1.2 and cipher suites ECC and RSA as standard. In addition, when a high level of security is required, XML signatures are to be used.

8.1.2 Communication Services According to OpenADR Standard

OpenADR 2.0 is formulated as part of the US energy-related standard **EI 1.0 (Energy Interoperation 1.0)** [3]. In EI 1.0, the various communications (services) that make up the demand response as an OpenADR profile are organized into the eight shown in Table 8.2. And OpenADR 2.0 has already been realized with four of these services.

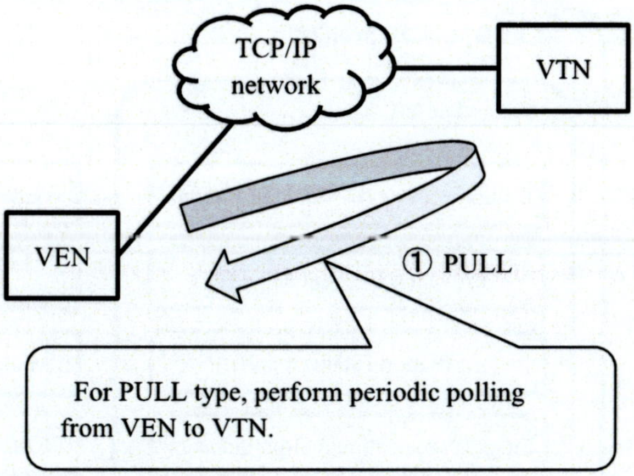

Fig. 8.2 OpenADR PULL communication sequence

Table 8.2 Services included in EI 1.0 OpenADR profile

Service	Meaning
EiRegisterParty	A service for registering, updating, and deleting VEN's own information for VTN
EiEvent	A service that VTN notifies/changes/cancels demand response events to VEN. This demand response event includes start/end time and price information at that time
EiReport	A service for when VEN notifies the VTN of the status of the equipment targeted for demand response or requests it from the VTN to VEN
EiOpt	A service for VEN to notify the VTN of changes in the availability of short-term participation in demand response events and acceptance/rejection of demand response events
EiEnroll[a]	A service that enables VEN to participate in demand response for VTNs
EiAvail[a]	A service for VEN to notify the VTN of the availability of long-term participation in demand response events
EiMarketContext[a]	A service that provides information on the electricity market and its demand response programs
EiQueue[a]	A service that provides electricity price information

Service names with [a] are not included in the current OpenADR 2.0

This is for the purpose of realizing the demand response services already provided at the time of formulation, and the remaining four services are to be included in future specifications. This is summarized in Fig. 8.3.

Figure 8.4 shows an example of demand response realized by actually using these services. This example shows registration of VEN using XMPP as the communication method, status reporting of demand response target equipment, and response and acceptance of demand response events. In OpenADR, each service is realized by exchanging several payloads (message contents) according to a predetermined exchange sequence.

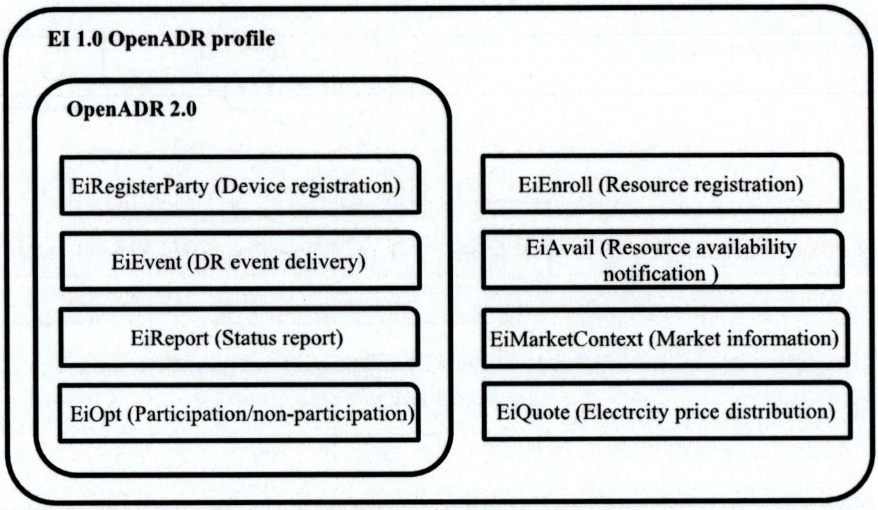

Fig. 8.3 Services that make up OpenADR

Next, among the services that make up OpenADR, the payload and exchange sequence of the most important EiEvent service are briefly described. The EiEvent service is a service for VTN to notify, change, and cancel demand response events to VEN. There are two main payloads that includes in this service: oadrDistributeEvent and oadrCreatedEvent. Specifically, demand response event information is transmitted using the oadrDistributeEvent payload. The oadrCreatedEvent payload is used to respond as acknoldgement of oadrDistributedEvent to VTN indicating whether or not a demand response event participated.

Figure 8.5 shows the exchange sequence in EiEvent service for both inquery method. The oadrDistibuteEvent payload is sent directly from the VTN to the VEN with PUSH mode, whereas it is sent as a response to the notification request (polling) payload that the VEN periodically sends to the VTN for the PULL mode.

It can also be obtained by sending the oadrRequestEvent payload from VEN to the VTN. The oadrCreatedEvent payload is sent asynchronously from VEN to VTN when requested from VTN (regardless of PUSH/PULL mode).

The items (XML elements) included in the oadrDistibuteEvent payload include the message ID (requestID element), the transmitted VTN ID (vtnID element), and zero or more demand response events (oadrEvent element). The oadrEvent element indicating an individual demand response event includes details of the event. An eiEvent element including the values of several demand response signals (such as price and usage change requests) and the period of the signals. There are two oadrResponseRequired elements that indicate whether VEN requires a response with oadrCreatedEvent payload or not.

The eiActivePeriod element of the eiEvent element defines the start time and execution time of the demand response event. The eiEventSignals element of

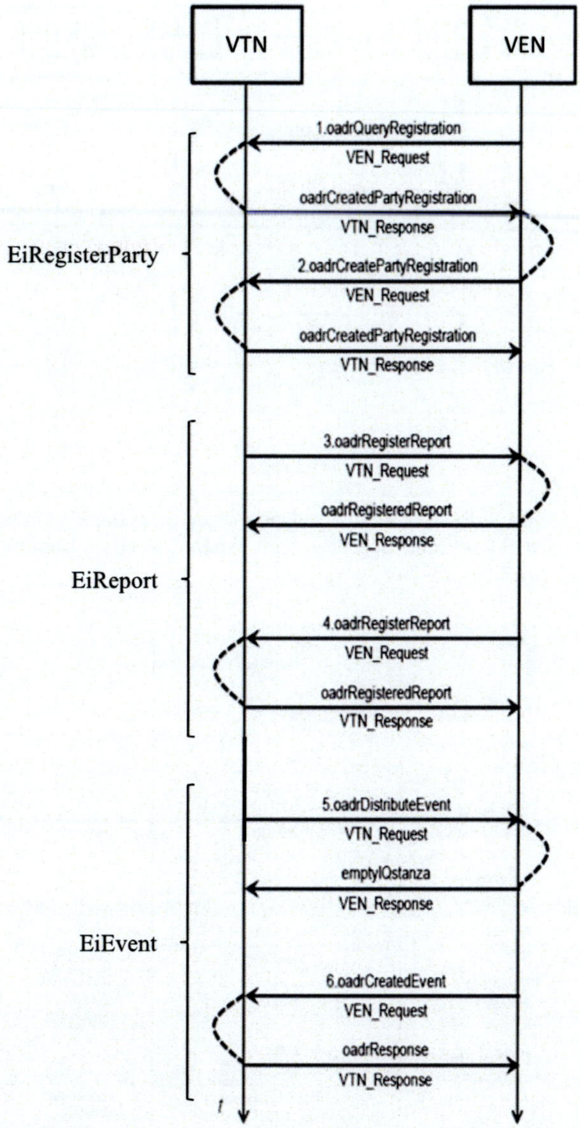

Fig. 8.4 Example of OpenADR demand response service communication

the eiEvent element has multiple eiEventSignal elements, and each eiEventSignal element defines the value and duration of the demand response signal that is executed in the demand response event. Figure 8.6 depicts the different time periods of an event.

The types of demand response signals specified differ depending on the profile. In 2.0a, only simple level control signal is specified for the eiEventSignal element. In 2.0b, allows for a more diverse set of signals in the event messages, in addition to the simple level of 2.0a, such as energy price, demand charge, bid price etc.

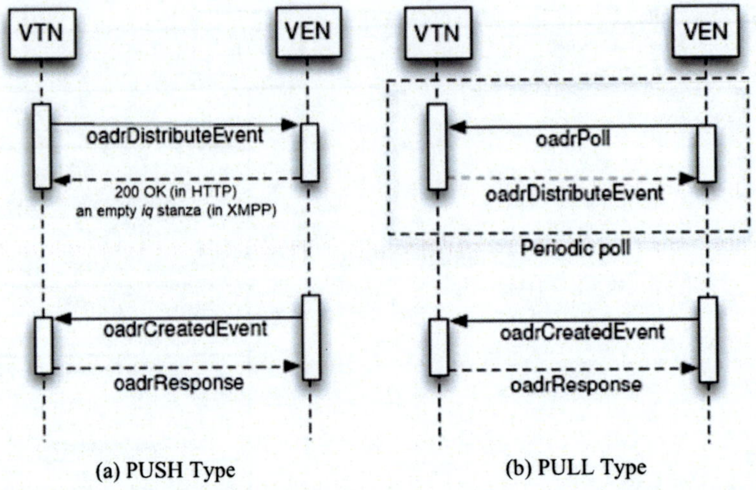

Fig. 8.5 EiEvent service exchange sequence example (Modified from OpenADR Alliance, "OpenADR 2.0 Profile Specification A Profile", Revision Number 1.0, www.openadr.org/specification (2015), pp. 23–24)

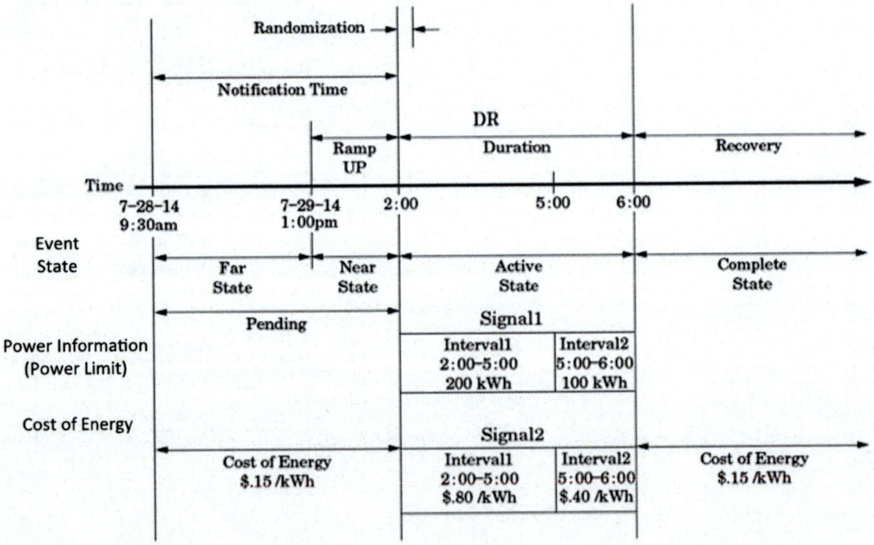

Fig. 8.6 Time intervals of an OpenADR event

In 2.0b, the payload used for polling is called oadrPoll. oadrPoll operation is used when the PULL mode inquery method is adopted. This oadrPoll is used to receive payloads that the VTN needs to notify. In other words, oadrPoll is periodically sent from the VEN to the VTN. The VTN may reply with an oadrDistributeEvent containing new or modified events. In 2.0a, polling is performed using the oadrRequestEvent payload instead of the oadrPoll payload.

8.1.3 OpenADR Standard Software Implementation

So far, the outline and architecture of OpenADR have been described. In this section, we will explain the implementation of OpenADR and introduce software that can actually confirm the operation.

① Difference between OpenADR 2.0a and 2.0b

As explained earlier, OpenADR has two profiles, 2.0a and 2.0b. This section details these differences.

Since OpenADR 2.0a targets simple devices, only EiEvent is included in the communication service, and only four-level control from 0 to 3 is supported. There are some items that are not supported. In addition, only Simple HTTP is required as a communication method, XMPP is optional, and the endpoint address is also different from 2.0b (Fig. 8.7).

In OpenADR 2.0b, it is mandatory to implement all four services mentioned above as communication services. Of course, there are no restrictions on EiEvent control. As for the communication method, both Simple HTTP and XMPP are mandatory for VTN, and either one of them is mandatory for VEN.

On the other hand, as VTN and VEN implementation requirements, VTN must support both 2.0a and 2.0b, and VEN requires either one. Note that until OpenADR 2.0b was formulated and released (of course), VTN was allowed only for 2.0a implementation. So when selecting VTN software etc., be careful whether 2.0b is supported or not.

The famous OpenADR open source implementations are summarized in Table 8.4. As can be seen from this table, it can be divided into two categories depending on the provider. Therefore, the details of the implementation by each provider are explained subsequently. Table 8.3 summarizes the services implementation rules in OpenADR.

② Acquire Open DR System

If the goal is simply to build an OpenADR system, the first thing to consider is to use an OpenADR-certified product.

On the other hand, when developing a new VTN/VEN in OpenADR system, an open source implementation may be used as an option. In general, it can be used free of charge.

```
<?xml version="1.0" encoding="UTF-8" standalone="no"?>
<iq from="vtn@www.gifu-u.ac.jp/EiEvent" id="Lf6gp-24"
  to="ven_java@www.gifu-u.ac.jp/ven" type="set">
<oadrSignedObject xmlns="http://openadr.org/oadr-2.0b/2012/07"
  :
 <oadrDistributeEvent ei:schemaVersion="2.0b">
   :
  <ei:vtnID>EPRI_VTN</ei:vtnID>
  <oadrEvent>
   <ei:eiEvent>
     :
    <ei:eiActivePeriod>
     <ical:properties>
      <ical:dtstart>
       <ical:date-time>2019-05-01T03:00:00.000Z</ical:date-time>
      </ical:dtstart>
      <ical:duration>
       <ical:duration>PT15M</ical:duration>
      </ical:duration>
     </ical:properties>
    <ei:eiActivePeriod>
     :
    <ei:eiEventSignals>
     <ei:eiEventSignal>
      <strm:intervals>
      <ei:interval>
       <ical:duration>
        <ical:duration>PT15M</ical:duration>
       </ical:duration>
       <ei:signalPayload>
        <ei:payloadFloat>
         <ei:value>20.0</ei:value>
        </ei:payloadFloat>
       </ei:signalPayload>
         :
```


XMPP OpenADR

Fig. 8.7 Payload example of OpenADR using XMPP communication method

③ OpenADR Open Source Implementation by EnerNoc, Inc.

EnerNoc, Inc. is the world's largest demand response company headquartered in the United States and a member of Enel Group, a leading Italian energy company. In 2013, Enanoc Japan Co., Ltd. was established as a joint venture with Marubeni Corporation in Japan.

There are two open source implementations [4] of VTN and VEN in OpenADR released by EnerNoc from 2013 to 2014.

(a) EnerNOC OpenADR 2.0 VTN
(b) EnerNOC OpenADR 2.0 Reference Implementation.

Implementation is done using Java and Groovy (a language that runs on the JVM), and the license is Apache License Version 2.0. Since VEN is only provided in the form of a test code, development is essential for actual use. In VEN, Simple HTTP is not implemented, but Simple XMPP is implemented, which is an advantage over the EPRI side implementation.

Table 8.3 Service implementation rules in OpenADR

	VTN		VEN	
	B	A	B	B (Energy reporting only)
Services and function support				
EiEvent				
Limited profile (2.0a specification)	M	M	NA	NA
Full profile	M	NA	M	NA
EiOpt				
Full profile	M	NA	M	NA
EiReport				
Full profile	M	NA	M*	M*
EiRegister party				
Full profile	M	NA	M	M
Transport protocols				
Simple HTTP	M	M	O-1	O-1
XMPP	M	NA	O-1	O-1
Security Levels				
Standard	M	M	M	M
High	O	NA	O	O

M Mandatory *NA* Not available for profile
O Optional *: Optional features available
O-1: Optional, but at least one of them must be supported

Table 8.4 Service implementation rules in OpenADR

Name	Provider	Type	Communication		License	Language/framework
			Simple HTTP	Simple XMPP		
EnerNOC OpenADR 2.0 VTN	EnterNoc	VTN	○	○	Apache License Version 2.0	Groovy, Grails
EnerNOC OpenADR 2.0 Reference Implementation		VEN	–	○	Apache License Version 2.0 (Includes someGPLv2 Code with classpath exceptions)	Java
OpenADR Virtual Top Node	EPRI	VTN	○	○	BDS 3-Clause license	JRuby, Ruby on Rails
OpenADR Virtual End Node		VEN	○	–	BDS 3-Clause License	C#

④ OpenADR Open Source Implementation by EPRI

EPRI (Electric Power Research Institute, Inc.) is a power research institute established in the United States in 1972 (equivalent to the Central Research Institute of Electric Power in Japan). It is an independent non-profit organization. Companies from all over the world, including Japan as well as the United States, participate as members.

The open source implementation [5] of VTN and VEN of OpenADR provided by EPRI was released from 2016 to 2017, and is shown below.

(a) OpenADR Virtual Top Node
(b) OpenADR Virtual End Node.

VTN is implemented in JRuby (Ruby running on JVM) and can run on Ubuntu 16.04. In contrast, VEN is implemented in C# and can be run on Windows 7 and above. Both licenses are BSD 3-Clause license.

In addition, what EPRI provides has been certified by OpenADR and is also introduced on the Alliance Web page.

⑤ OpenADR Authentication Procedure

When developing and selling OpenADR-compatible products, the following procedures are required to be certified as compliant products based on the OpenADR Alliance certification program.

(a) Membership in the OpenADR Alliance
(b) Completion of protocol implementation conformity statement (PICS)
(c) Completion of certification test based on PICS by a testing institution accredited by the OpenADR Alliance
(d) Submit required documents to the OpenADR Alliance
(e) Approval of OpenADR Alliance.

An example of EiEvent service is shown as a specific example of OpenADR. In this example, the power command value $P_L = 20.0$ kW is set for 15 min from 3:00 (UTC) on May 1, 2019.

This example is a payload sent by XMPP communication, and an OpenADR message is loaded in the XMPP data part. In HTTP communication, the XMPP header part is replaced with the HTTP header, and the OpenADR message in the data part is common.

8.2 IEC 61850 Communication Standard

8.2.1 IEC 61850 Logical Nodes Configuration

All functions required for substation automation applications (e.g. protection, control, measurement, monitoring, etc.) need to be identified by job role, start criteria, expected output, defined performance criteria. These functions are broken down into logical nodes.

The general approach of the IEC 61850 series is to break down application functions into the smallest entities and use them to exchange information as described in IEC 61850-5. Granularity is given by rationally distributing these entities on dedicated devices (IEDs). This entity is a logical node. The requirements for logical nodes are defined in IEC 61850-5 from the application perspective.

These logical nodes contain data and data attributes based on their function. Information represented by data and data attributes is exchanged by dedicated services according to clear rules and functions required by IEC 61850-5.

Each logical device contains one or more logical nodes. A logical node is a named group of some logically related data and related services.

The physical model of IEC 61850 starts with a physical device. A physical device is a device connected to a network. A physical device is usually defined by its network address. There may be one or more logical devices in each physical device. The IEC 61850 logical device model allows a single physical device to function as a proxy or gateway for multiple devices, providing a standard representation of a data concentrator.

Figure 8.8 shows the concept and data structure of IEC 61850. Logical nodes are grouped according to the logical node group shown in the left column of Table 8.5. The name of a logical node begins with an identifier acronym that represents the group to which the logical node belongs. Since the modeling for each phase (for example, circuit breaker) is functionally a switch, it is classified as "Switchgear"

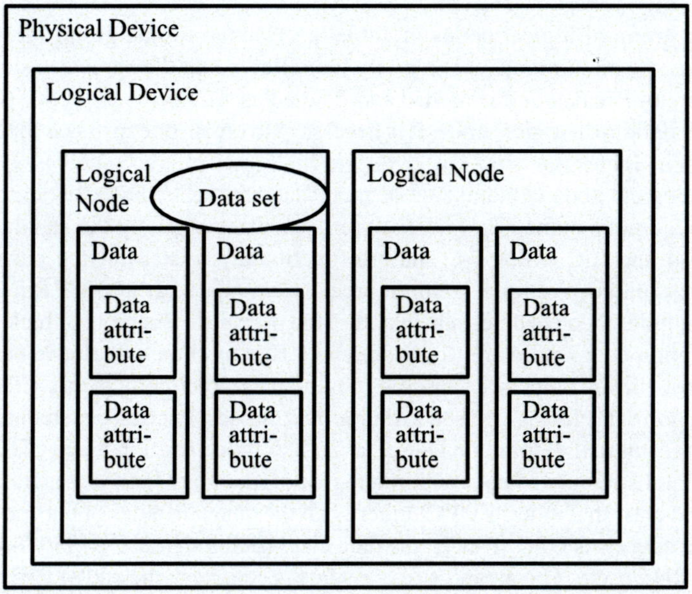

Fig. 8.8 IEC 61850 concept and data structure

Table 8.5 Logical node groups (Quoted from IEC 61850-7-4 (2003), p. 15)

Group indicator	Logical node groups
A	Automatic control
C	Supervisory control
G	Generic function references
I	Interfacing and archiving
L	System logical nodes
M	Metering and measurement
P	Protection functions
R	Protection related functions
S	Sensors, monitoring
T	Instrument transformer
X	Switchgear
Y	Power transformer and related functions
Z	Further (power system) equipment

Table 8.6 Example of naming logical nodes

Name of logical node	Meaning
XCBR1	Short Circuit Breaker unit 1
XCBR2	Short Circuit Breaker unit 2
XCBR3	Short Circuit Breaker unit 3

from the group of logical nodes in Table 8.5. Therefore, the group identifier is X. Therefore, the circuit breaker is a group identifier X and the name is a short form of Short Circuit Breaker and is named and defined as XCBR.

When defining a logical node, it is necessary to create one instance for each phase as shown in Tables 8.6 and 8.7.

Each logical node contains one or more data elements. Each key element in the data has a unique name. These data names are determined by the standards of IEC 61850-7-4 and IEC 61850-7-3 and are functionally related for the purpose of the power system. For example, a circuit breaker is modeled as an XCBR logical node, which includes a Loc to determine whether the operation is remote or local, an OpCnt for operation count, a Pos for location, and a BlkOpn block. Includes breaker open command, BlkCls block breaker close command, and circuit breaker CBOpCap.

Common data classes are used to define data classes. Each data element in a logical node conforms to the Common Data Class (CDC) specification. Each CDC describes the type and structure of data within a logical node. For example, the data class Pos (a derived class that inherits DPC) inherits all the data attributes of the corresponding common data class DPC. That is, the data class (defined in IEC 61850-7-4) is ctlVal, origin, ctlNum, etc. The meaning of class Pos is defined at the end of IEC 61850-7-4.

Table 8.7 Logical node groups (Quoted from IEC 61850-7-4 (2003), p. 15)

XCBR class				
Attribute name	Attr. type	Explanation	T	M/O
LNName		Shall be inherited from Logical-Node Class (see IEC 61850-7-2)		M
Data				
Common logical node information				
		LN shall inherit all Mandatory Data from Common Logical Node Class		M
Loc	SPS	Local operation (local means without substation automation communication, hardwired direct control)		M
EEHealth	INS	External equipment health		O
EEName	DPL	External equipment name plate		O
OpCnt	INS	Operation counter		M
Controls				
Pos	DPC	Swith positon		M
BlkOpn	SPC	Block opening		M
BlkCls				
ChaMotEna	SPC	Charger motor enabled		O
Metered Values				
SumSwARs	BCR	Sum of Switched Amperes, resetable		M
Status Information				
CBOpCap	INS	Circuit breaker operating capability		M
POWCap	INS	Point On wave switching capability		O
MaxOpCap	INS	Circuit breaker operating capability when fully charged		O

A storage battery system is generally configured as shown in Fig. 8.9. The storage battery system consists of a storage battery rack to store the storage battery, a DC/AC conversion power conditioner PCS (Power Conditioning System), and a storage battery control system. The storage battery system is connected to the in-facility power system through a grid connection breaker, but a step-up transformer is used depending on the voltage classification of the in-facility power system.

Apply the information model of IEC 61850 to the storage battery system. Since the entire storage battery system is a physical device and can be called a single logical device connected to a network, it can be defined as IED. There are one circuit breaker

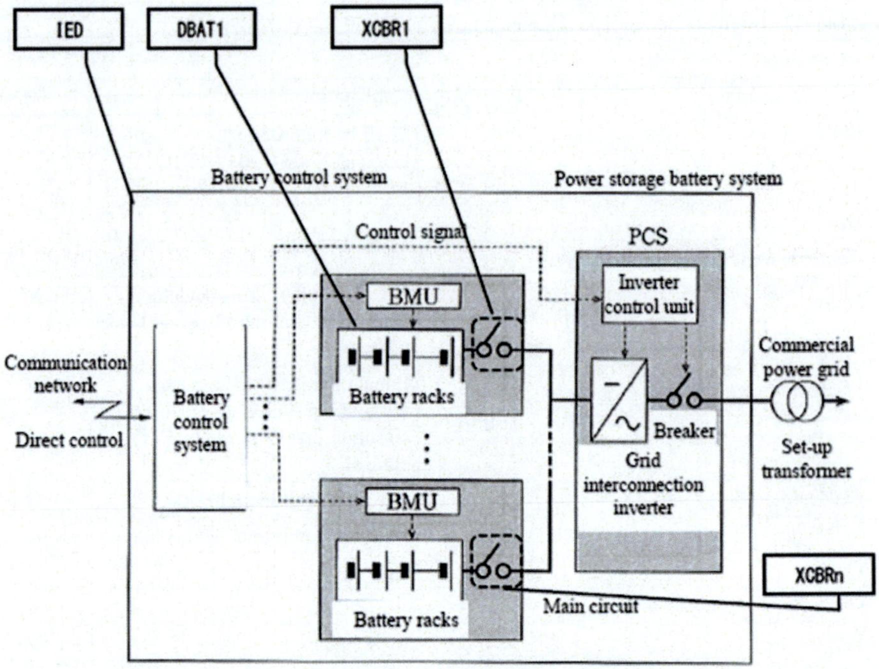

Fig. 8.9 General configuration of storage battery system (power storage storage)

in the PCS and one in each of the three battery racks. Each can be defined as a logical node $XCBR_1$ ... $XCBR_n$. Devices such as inverters, inverter control units, storage battery control systems, storage batteries, and BMUs can also be considered as the smallest entity of application functions, that is, a logical node. These entities are defined as logical nodes in IEC 61850-7-4.

8.2.2 Implementation of IEC 61850 Standard Software

IEC 61850 was initially standardized for the purpose of exchanging information between intelligent electronics devices (IEDs) in substation automation systems, but is now extensive and comprehensive on communication networks and systems for power utility automation. It has become a standard and is being extended for the smart grid in the future. Therefore, the description of substations is sometimes found in the standard document, but that is the background, so there is no problem replacing substations with storage batteries.

As explained in Chap. 3, the IEC 61850 series has a hierarchical structure as shown in Fig. 8.10. The information model is a logical node for exchanging information from the top of the hierarchy, a service interface that defines how information is exchanged,

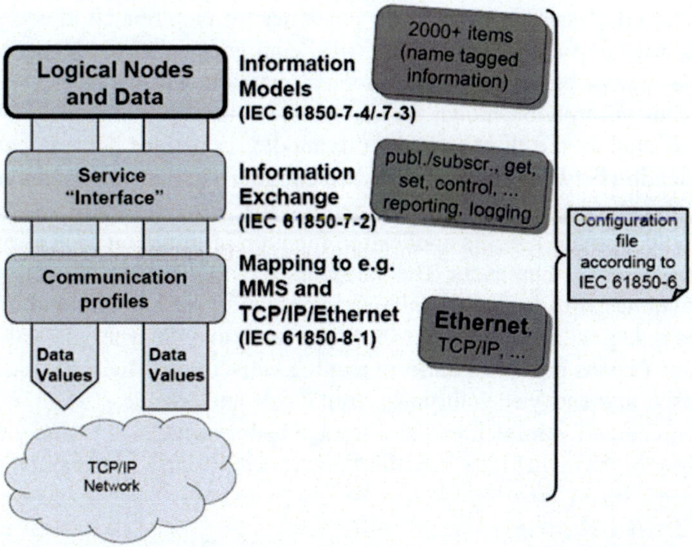

Fig. 8.10 Outline of IEC 61850 series concept (Quoted from IEC 61850-7-1 (2003), p. 24)

and a communication profile that actually conveys information. SCL (Substation Configuration description Language) [6] is defined so that these can be "configured" by XML description. The IEC 61850 series consists of an enormous number of standard documents, and each standard is explained in detail in its respective scope. Therefore, it is difficult to understand the relationship between standards. Table 8.9 shows the main standard documents related to the battery system selected by the author.

A brief introduction to the standard text gives an overview of the use of IEC 61850 from a technical point of view, such as system design, in Part 7-1.

The logical node (Logical Node) that is the core of the information model in the upper left of Fig. 8.10 is defined in Part 7-3 and Part 7-4. In Part 7-3, common data for controlling and measuring electrical equipment is used. A basic information model called class, and a model for exchanging information are defined in Part 7-4.

The service interface (ACSI: Abstract Communication Service Interface) [7], which is the information exchange model in the middle left, is defined in Part 7-2. This service model is an abstract form and is separated from communication protocols and communication services.

The communication profile, which is an information exchange model in the lower left in Fig. 8.10, is defined separately for each specific communication service mapping (SCSM) [8]. Part 8-1 defines MMS (ISO 9506-1 and ISO 9506-2) and ISO/IEC 8802-3 mapping, Part 9-2 specifies SV (Sampled Value) and ISO/IEC 8802-3 mapping. SCL, the IED configuration description language, is specified in Part 6.

"Logical node" can be said to be a container for data related to equipment and functions related to automation of electrical equipment. IEC 61850 defines a basic conceptual model based on the hierarchical class shown in Fig. 8.11 in order to construct an information model for electrical equipment automation. The logical node is located in the third right of this model, is defined for compatibility for communication between IEDs, and plays a central role in the information model.

The logical node class defines functions that provide application functions. For example, domain-specific application function information such as overvoltage protection and circuit breakers. The data class provides a means to define the location of switches with typed information, e.g. quality information and time stamps contained in logical nodes. Data attribute classes define the types and values used in data. Name classes provide a uniform naming convention. The functionality of this class is used to identify each instance from inside and outside.

The procedure for implementing a storage battery system is to apply the storage battery system shown in Fig. 8.12 to this basic conceptual model. Here, as an example,

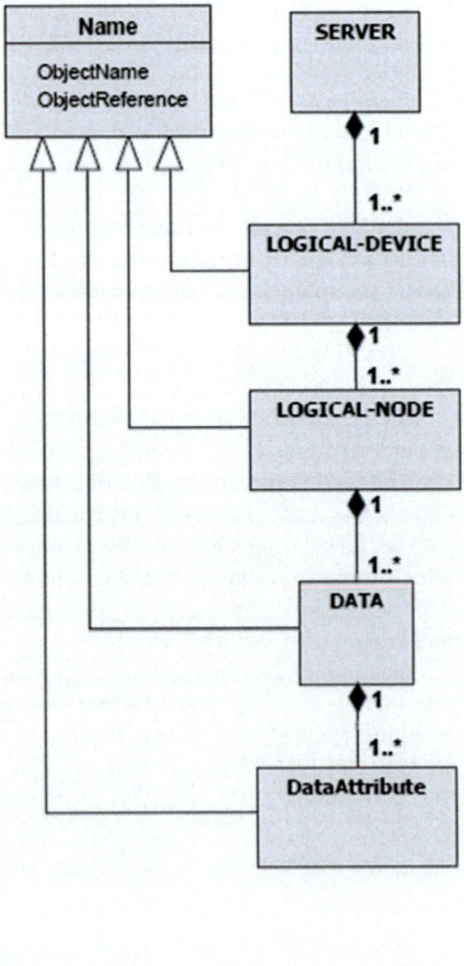

Fig. 8.11 Basic conceptual model of IEC: ACSI

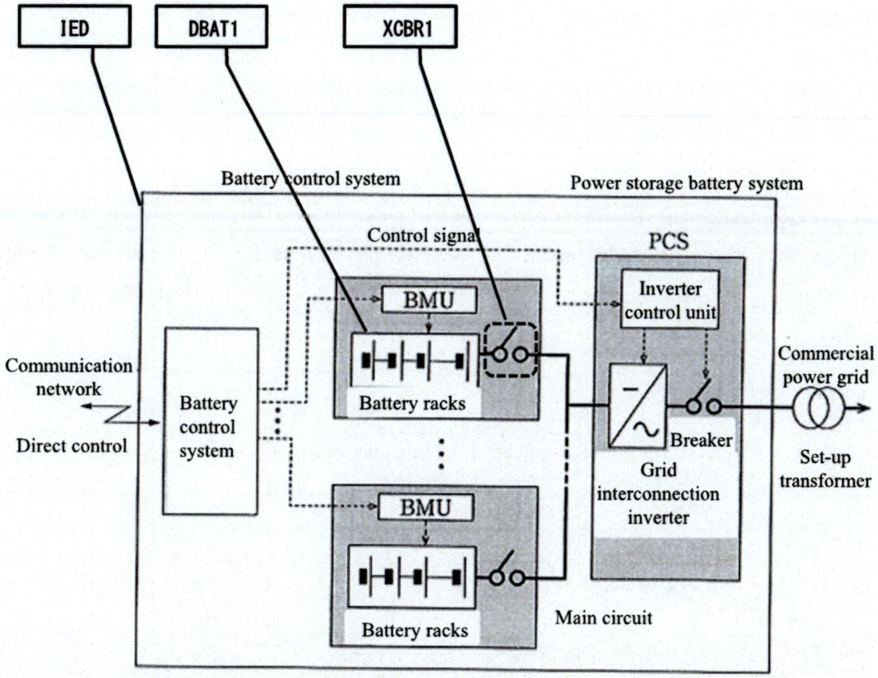

Fig. 8.12 Storage battery system model using logical nodes

the entire storage battery system is defined as a logical node called IED, one circuit breaker as XCBR, and one storage battery as DBAT. The logical node XCBR is specified in Part 7-4 and DBAT is specified in Part 90-9. The classes in Tables 8.8 and 8.10, respectively, are defined as logical nodes that provide application functions (Table 8.9).

When the basic conceptual model is applied to a storage battery system, the XCBR logical node instance is XCBR1 and the DBAT logical node instance is DBAT1, the storage battery system class is as shown in Fig. 8.13.

The server class represents the operation that can be seen from outside the target electrical equipment. Therefore server under the class, one storage battery system is defined as the IED class of the logic device.

Under the logical device IED class, logical nodes that serve as application functions are placed. Define an instance XCBR1 of the circuit breaker XCBR class and an instance DBAT1 of the storage battery DBAT class.

The data is arranged under the logical node DBAT1 with reference to the data object name and common data class of Table 8.10 DBAT logical node. Depending on the logical class, optional, and conditional data classes are required. Select and define the necessary data. A data attribute class is defined under the data class.

For example, the SOC value indicating the state of charge of the storage battery is specified so that the instance of the SocPct data class under DBAT1 holds the value

Table 8.8 Logical node groups (Quoted from IEC 61850-7-4 (2003), p. 15)

DPC class					
Attribute name	Attribute type	FC	TrgOp	Value/value range	M/O/C
DataName	Inherited from data class (see IEC 61850-7-2)				
DataAttribute					
Control and status					
cltVal	BOOLEAN	CO		off (FALSE) I on (TRUE)	AC_CO_M
operTm	TimeStamp	CO			AC_CO_O
origin	Originator	CO, ST			AC_CO_O
cltNum	INT8U	CO, ST		0.0.255	AC_CO_O
stVal	CODED ENUM	ST	dchg	intermediate-state I off I on I bas-state	M
q	Quality	ST	qchg		M
t	TimeStamp	ST			M
stSeld	BOOLEAN	ST	dchg		AC_CO_O
Substitution					
subEna	BOOLEAN	SV			
subVal	CODED ENUM	SV		intermediate-stateIofflonIbas-state	
subQ	Quality	SV			
subID	VISIBLE STRING64	SV			
Configuration, description and extension					
pulseConfig	PulseConfig	CF			AC_CO_O
ctlModel	CtlModels	CF			M
sboTimeout	INT32U	CF			AC_CO_O
sboClass	SboClasses	CF			AC_CO_O
d	VISIBLE STRING255	DC		Text	O
dU	UINCODE STRING255	DC			O
cdcNs	VISIBLE STRING255	EX			AC_DLNDA_M
cdcName	VISIBLE STRING255	EX			AC_DLNDA_M
dataNs	VISIBLE STRING255	EX			AC_DLN_M
Services					
As defined in Table 31 of IEC 61850-90-9					

Table 8.9 Standard documents for IEC 61850 series related to storage battery systems

IEC 61850-6	Configuration description language for communication related to IED
IEC 61850-7-1	Basic communication structure-principles and models
IEC 61850-7-2	Basic communication structure-Abstract Communication Service Interface (ACSI)
IEC 61850-7-3	Basic communication structure-common data class
IEC 61850-7-4	Basic communication structure-compatible logical node class and data object class
IEC 61850-7-420	Basic communication structure-distributed energy resource logical node
IEC 61850-7-500	Basic information and communication structure-substation automation system logic node
IEC 61850-7-520	Distributed power logical node
IEC 61850-8-1	Specific Communication Service Mapping (SCSM)-Mapping to MMS (ISO 9506-1 and ISO 9506-2) and ISO/IEC 8802-3
IEC 61850-9-2	Specific Communication Service Mapping (SCSM)-value sampled by ISO/IEC 8802-3
IEC 61850-90-9	IEC 61850 for storage battery systems

indicating the percentage of the SOC as an integer type or decimal point type of the data attribute class.

"Service interface" provides abstract definitions of classes and services so that they do not depend on a specific protocol stack, implementation, or operation system. IEC 61850 aims to provide interoperability between electrical facilities, and communication between IEDs is a hierarchical class model (logical device, logical node, data, data set, report control, log, etc.) of the basic conceptual model.) And services provided by these classes (get, set, report, etc.).

IEC 61850 defines an Abstract Communication Service Interface (ACSI) for use in domains that require real-time coordination of IEDs. ACSI is defined independently of the underlying communication system. In addition, specific communication service mapping (SCSM) is specified in Part 8 and Part 9, and the hierarchical class model of all information accessible via the communication network, the services operating in these classes, and the parameters associated with each service are specified.

The services that the basic conceptual model class in Fig. 8.11 has are shown in Table 8.11. For example, the GetServerDirectory service in the server class provides a function to search all logical devices or file instances under the server and obtain a list of all names for the accessing client Fig. 8.14. The GetDataValues service of the data class provides the function to get the value of the specified data attribute of the referenced data, and the SetDataValues service provides the function to set the value of the data attribute in the same way. Dataset classes define groupings of data and data attributes. Used for direct access and report and log classes.

Each class of the hierarchical class model (logical device, logical node, data, data attribute) of the basic conceptual model inherits a common name class, and the

Table 8.10 IEC: Logical node of DBAT battery system (Partially modified from IEC 61850-90-9 (2018), p. 54)

DBAT				
Data object name	Common data class	T	Explanation	Prerequisite conditions
Descriptions				
EEName	DPL		Inherited from: ControlEquipmentInterfaceLN	O/F
NamPlt	LPL		inherited from: DomainLN	MONamPlt/MONamPlt
Status information				
BatSt	ENS (ENSBatteryState)		State of the battery	M/F
ChaSt	SPS		If true, battery is charging	O/F
ExtTmpLoAlm	SPS		If true, the high level threshold of the temperature external to the battery has been reached	M/F
Measured and metered values				
Amp	MV		Battery discharging DC current	O/O
Watt	MV		DC power of storage battery	O/O
ChaAmpLim	MV		Instantaneous DC current limit on charging	O/O
ChaVolLim	MV		Instantaneous voltage limit on charging	O/O
DschAmpLim	MV		Instantaneous DC current limit on discharging	O/O
IntnTmp	MV		Internal battery temperature in degrees C	O/O
ExtTmp	MV		External battery temperature in degrees C	O/O
ExtVol	MV		External battery voltage	O/O
IntnVol	MV		Internal battery voltage	O/O
CelStrgCnt	MV		Number of strings of cells currently connected	O/O
VolChgRte	MV		Rate of battery voltage change over time	O/O
DschVolLim	MV		Instantaneous voltage limit on discharging	O/O
SocPct	MV		Storage battery SOC [%]	O/O
:				

M Mandatory *O* Optional *F* Forbidden

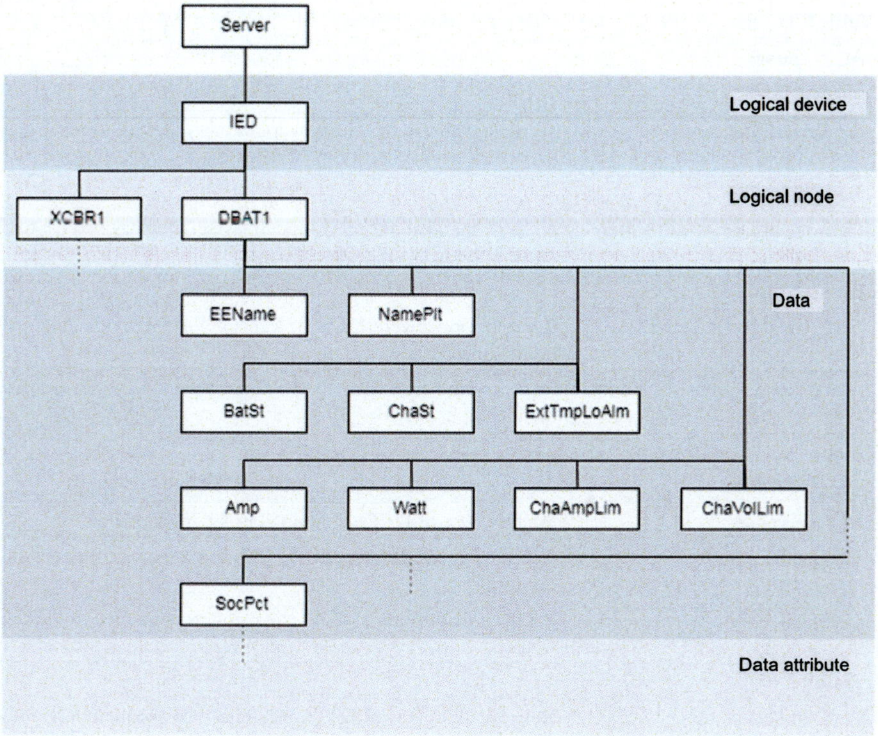

Fig. 8.13 Storage battery system class structure according to IEC 61850

instance of each class is uniquely identified and accessed using this name class to enable.

The ObjectName data attribute class is defined as a string type of up to 32 characters by the ACSI common data class, and provides a function to identify a unique instance name among instances of the class owned by the same parent class. The ObjectReference data attribute class is defined as a string type of up to 255 characters, and provides a complete path function concatenating unique instance names of the hierarchy information class. In the path indicating the hierarchy, the logical device name and logical node name are distinguished by "/", and the subsequent hierarchy is separated by ".". [] Indicates an option.

Logical device name/Logical node name [.data name [.data attribute name [. ...]]].

The "Communication Profile" specifies a method for sending and receiving data using a specific communication medium and communication protocol via a "Service Interface" and Specific Communication Service Mapping (SCSM). Figure 8.15 shows the basic communication model of IEC 61850. The information model of IEC 61850 is at the top of the Fig. 8.15, and the three arrows pointing downward indicate the selection of the communication method: client–server communication, GOOSE, and Sampled Values.

Table 8.11 IEC: ACSI service (Partially modified from IEC 61850-7-2 (2003), p. 19)

ACIS classes	
Server	
GetServerDirectory	A client shall use the GetServerDirectory service to retrieve a list of the names of all logical devices
Logical Device	
GetLogical Device Directory	A client shall use the GetLogicalDeviceDirectory service to retrieve the list of the ObjectReferences of all logical nodes
Logical Node	
GetLogical Node Directory	Retrieve ObjectReferences of a specific ACSI class contained in the logical nodes
GetALLDtatValues	Retrieve all the DataAttribute values of all DATA contained in the logical nodes
Data	
GetDataValues	Retrieve values of DATA contained in the logical node
SetDataValues	Write values of DATA contained in the logical node
GetDataDirectory	Retrieve ObjectReferences of all DataAttributes contained in the DATA
GetDataDefinition	Retrieve definitions of all DataAttributes contained in the DATA
DataSet	
GetDataSetValues	Retrieve all values of DATA referenced by the members of the DATA-SET
SetDataSetValues	Write values of DATA referenced by the members of the DATA-SET
CreateDataSet	Create a DATA-ST by providing the FCD (FCDA) references
DeleteDataSet	Delete a DATA-SET
GetDataDirectory	Retrieve FCD references of all members referenced in DATA-SET

Client–server communication is a client–server communication model using Ethernet and TCP/IP. Because TCP/IP is widely used in the world, it can be used regardless of the network type, such as LAN and Internet. In the case of IEC 61850, the IED is usually on the server side, and responds to requests from clients such as reading and writing data, controlling devices, and reporting when events occur. In client–server communication, the data of the information model is converted into data representation by MMS (Manufacturing Message Specification) in the mapping layer, encoded and structured by ASN.1 (Abstract Syntax Notation One), and then TCP/IP It is mounted on the payload (data) part of the packet and transmitted. The IEC 61850 client–server communication method stipulates that MMS and ASN.1 should be used, and this communication method is indicated as MMS in the standard text because it is more characteristic than client–server communication.

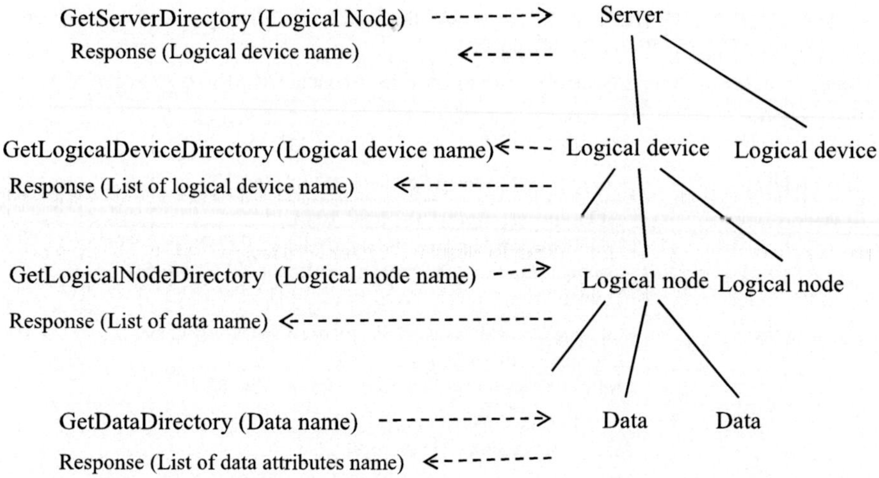

Fig. 8.14 IEC: GetDirectory service for each class (Partly modified from IEC 61850-7-2 (2003), p. 27)

Fig. 8.15 IEC: Overview of IEC 61850 communication model (Partially modified from IEC 61850-8-1 (2004), p. 19)

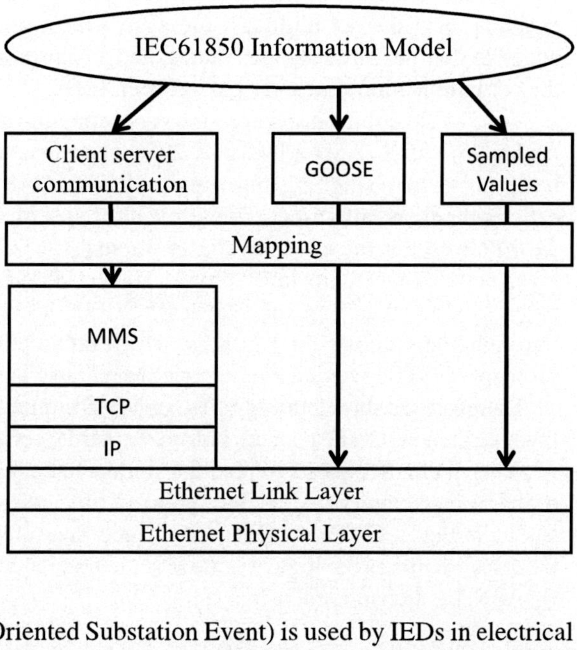

GOOSE (Generic Object Oriented Substation Event) is used by IEDs in electrical equipment to broadcast events. Instead of using TCP/IP, the data is directly loaded into the data part of the Ethernet frame and multicast is used. All kinds of process data can be sent between IEDs. Multicast is performed using the publish/subscribe (subscribe/subscribe) model.

Table 8.12 IEC: IEC 61850 MMS communication stack (TCP/IP) by OSI 7 layer (Partially modified from IEC 61850-8-1 (2004), pp. 23–24)

Application Layer	Association Control Service Element (ACSE)—ISO 8649/8650
Presentation layer	Connection Oriented Presentation—ISO 8855/8823 Abstract Syntax Notation (ASN)—ISO 8824/8825
Session layer	Connection Oriented Session—ISO 8326/8327
Transport layer	ISO transport over TCP—RFC 1006 ISO Transport Services on Top of the TCP (TPKT)—RFC 983 Connection Oriented Transport Protocol (COTP)—ISO 8073 Transmission Control Protocol (TCP)—RFC 793
Network layer	Internet Control Message Protocol (ICMP)—RFC 792 Internet Protocol (IP)—RFC 791 Address Resolution Protocol (ARP)—RFC 826
Link layer	IP datagrams over Ethernet—RFC 894 MAC—ISO 8802-3 [Ethernet]
Physical layer	Ethernet

Sampled Values (SV or SMV: Sampled Measured Values) are used for streaming live measured values obtained from sensors by IEDs in electrical equipment. For example, an Ethernet multicast message with a sampling rate of 4 kHz. Time is added to the measured sample values, and synchronous sampling is possible even if the communication time changes between IEDs.

Now, MMS [9] used in client–server communication is a specification standardized by ISO 9506, ASN.1 [10] is X.208 (revised version is X.680). Since IEC 61850 has a policy of using existing standards without defining communication specifications independently, this communication specification hierarchy is used. When TCP/IP is used in the lower layer as in Table 8.12. TCP/IP using Ethernet is transport layer, network layer, link layer, physical layer, MMS is application layer, presentation layer, session layer. Transport layer RFC 1006 encapsulates transport layer packets into individual units called TPKTs, which provide a connection-oriented transport protocol (COTP) over TCP to connect MMS and TCP/IP.

Therefore, as shown in Fig. 8.16, in MMS communication using TCP/IP, a lower layer sequence (COTP) occurs before the MMS communication sequence.

When a client requests to read the storage battery charge state SOC from the IED of the storage battery system, the syntax of the request message specified by MMS is shown in Fig. 8.17. The "MMS: Confirmed Read Request Sequence" in Fig. 8.17, a path that identifies the logical node is set in this request message, and data structured by ASN.1 is transmitted.

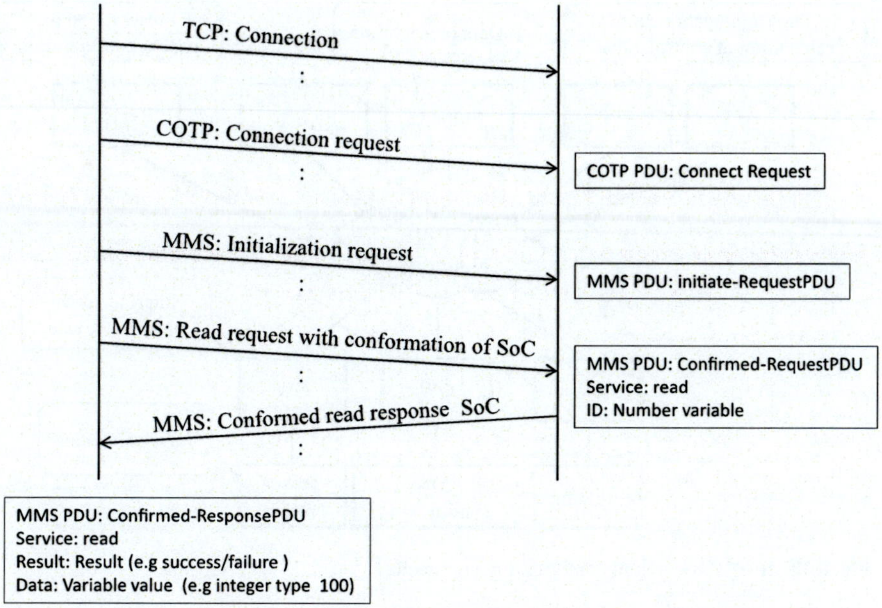

Fig. 8.16 MMS communication sequence for reading the SoC of the storage battery

```
Confirmed-RequestPDU ::= SEQUENCE {

        invokeID Unsigned32,
        listOfModifiers SEQUENCE OF Modifier OPTIONAL,
        service ConfirmedServiceRequest,
        ...,
        service-ext [79] Request-Detail OPTIONAL

                -- shall not be transmitted if value is the value
                -- of a tagged type derived from NULL

        }
```

Fig. 8.17 Read request code image with confirmation by MMS "Confirmed-RequestPDU" syntax (quoted from ISO 9506-2 (2003), p. 17)

8.3 IEEE1888 Communication Standard

8.3.1 Communication Method of IEEE 1888 Standard

The IEEE1888 system architecture consists of the four elements shown in Fig. 8.18.

- GW (gateway)

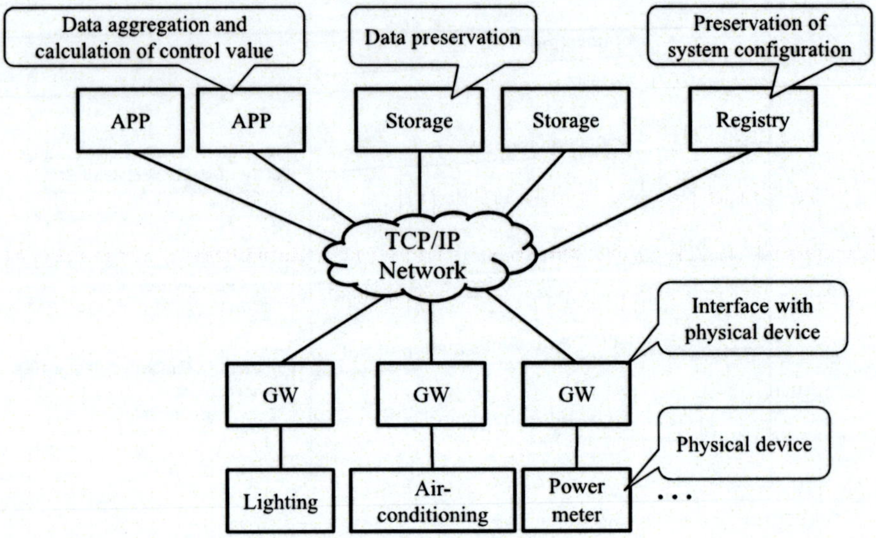

Fig. 8.18 IEEE1888 system configuration architecture

Connect physical devices such as lighting and air-conditioning to a TCP/IP network, and send and receive information such as sensor acquisition values and command values to and from Storage and APP using the IEEE1888 protocol.

- Storage
  Holds IEEE1888 data sent from APP or GW as a time-stamped database.
- APP
  Aggregate and graph data stored in Storage. In addition, the processing results are placed in Storage, and command values are sent to GW.
- Registry
  Maintains the configuration of the IEEE1888 system and responds to inquiries from GW, Storage, and Application.

In addition, GW, Storage, and APP are three components. Because Registry has different properties from components, different communication methods are defined. Registry is not essential for small systems with few components and only one storage. Conversely, in a large-scale system where a large number of components exist and data is distributed across multiple storages, the Registry plays an important role in system management.

In IEEE1888, monitoring control objects such as sensor measurement values of physical devices are managed by the concept of points. Figure 8.19 shows a conceptual diagram of points. Figure 8.19 shows an example when a GW writes a value measured by a physical device to Storage (WRITE). Point treats the value and time as a set. Values can hold integer values, decimal values, string values, and images. Note that the name of the point is named in the URL (Uniform Resource Locator) format, but it means the name used in the IEEE1888 system, not the actual

	Point a ID	UTC Time	Value
	⋮	⋮	⋮
	http://www.gifu-u.ac.jp/nlab/GW01/AC01/Power	2018-9-12 17:30:00+09	12.0
	http://www.gifu-u.ac.jp/nlab/GW01/AC01/Power	2018-9-12 17:31:00+09	13.0
Add to→	http://www.gifu-u.ac.jp/nlab/GW01/AC01/Power	2018-9-12 17:32:00+09	14.0

Storage holds time-series data for each point.

Fig. 8.19 Concept of monitoring control target in IEEE1888

Web page address is there. Storage can hold time-series data of each point and has a mechanism to refer from other APP and GW.

In IEEE1888, each component and registry communicate using HTTP. GW and APP have functions of Web server and/or Web client. On the other hand, Storage and Registry are Web servers in nature.

In HTTP, first, the Web client sends an HTTP request to the Web server. This HTTP request consists of a header and a body. Next, the Web server interprets the HTTP request sent from the Web client and returns an HTTP response to the Web client. This HTTP response consists of a header and a body.

In IEEE1888, communication is performed using a protocol called SOAP in which the body of the HTTP request and HTTP response is described in XML (Extensible Markup Language) format. Figure 8.20 shows a conceptual diagram of SOAP. SOAP is a protocol developed for calling processing and exchanging data between networks via different operating systems. SOAP is widely used as a means to implement Web services and is also applied to IEEE1888.

Each component has a method (function) called query and data, and the data (argument) passed to the method is described in the body to pass information and control the communication sequence. There are also registration and lookup methods in the registry, which are also used by describing them in the body. In this way, the mechanism for remotely invoking terminal processing is called remote procedure call (RPC), and SOAP.

Methods and data formats are described in a language called WSDL (Web Services Description Language), and each component and registry are made public so that they

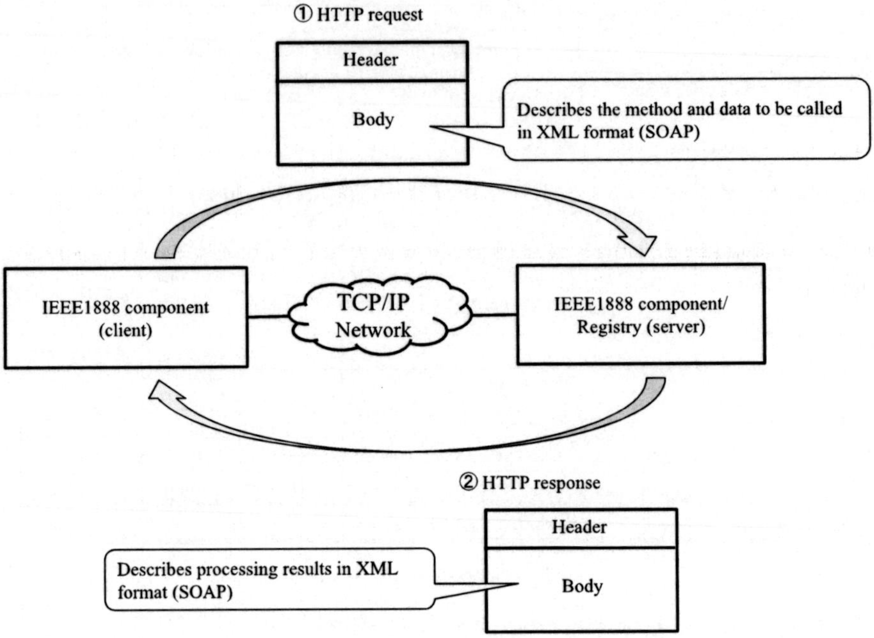

Fig. 8.20 Concept of communication by SOAP in IEEE1888

can understand each other's interface. This WSDL is also useful for software development using IEEE1888. Specifically, by loading the WSDL into an IDE (Integrated Development Environment), the source code describing the Web service interface is automatically generated, and the software. It can be used immediately as a part.

In IEEE1888, the following communication is realized by each method of data, query, registration, and lookup.

- WRITE
  Send point data to other components.
- FETCH
  Get the specified point data from other components.
- TRAP
  When there is a change in a point specified in advance, the registered component is notified of the change (WRITE).
- REGISTRATION
  Register the component name, Web address, supported communication method, validity period, etc. in Registry. Also, register the point ID, input/output distinction, period, etc. handled by the component.
- LOOKUP
  Specify the point ID, period, etc. for Registry and search for the component that owns the point.

For example, as shown in Fig. 8.21, the communication between components can be a system in which GW writes sensor information etc. to Storage and APP periodically FETCHs sensor information in Storage to graph it. At this time, the time interval for transmitting information from GW to APP depends on the polling interval of FETCH.

On the other hand, as shown in Fig. 8.22, if APP traps Storage, Storage can notify APP of the change.

Communication between components and registries is roughly divided into two types: point registration/search and component registration/search.

Regarding point registration/search, for example, in the system shown in Fig. 8.23, the GW newly connected to the system registers (REGISTRATION) the attributes (room number, value type, etc.) of the points handled by itself in the Registry. At this time, Registry holds the point table, and holds the point ID and its attributes. And APP can get the point name from the attribute information by LOOKUP to Registry.

On the other hand, for component registration/retrieval, for example, as shown in Fig. 8.24, the GW newly connected to the system is the component location (URI),

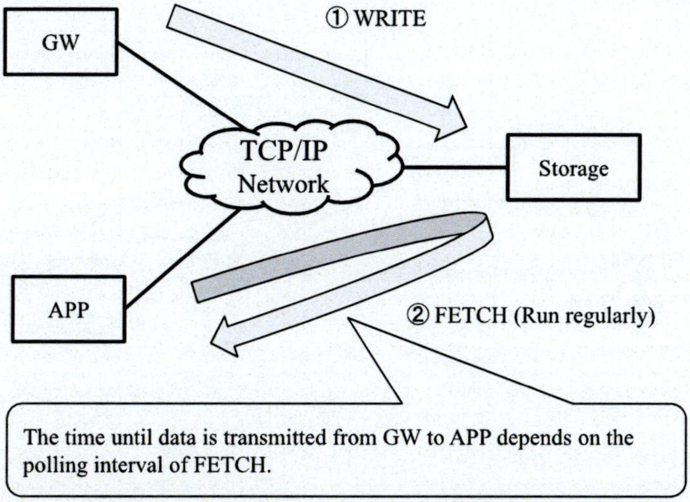

Fig. 8.21 Communication between components (WRITE, FETCH)

Fig. 8.22 Communication between components (TRAP, WRITE)

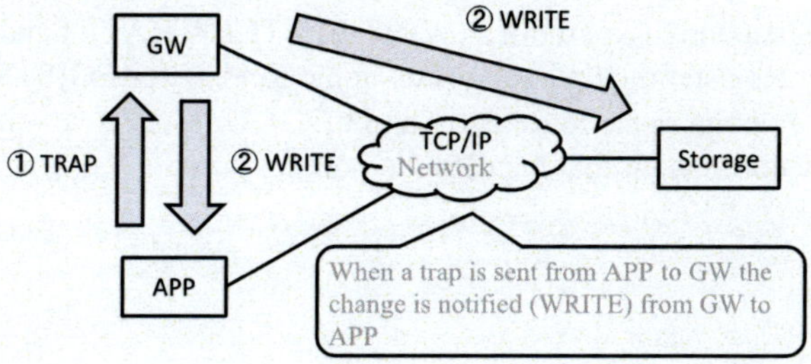

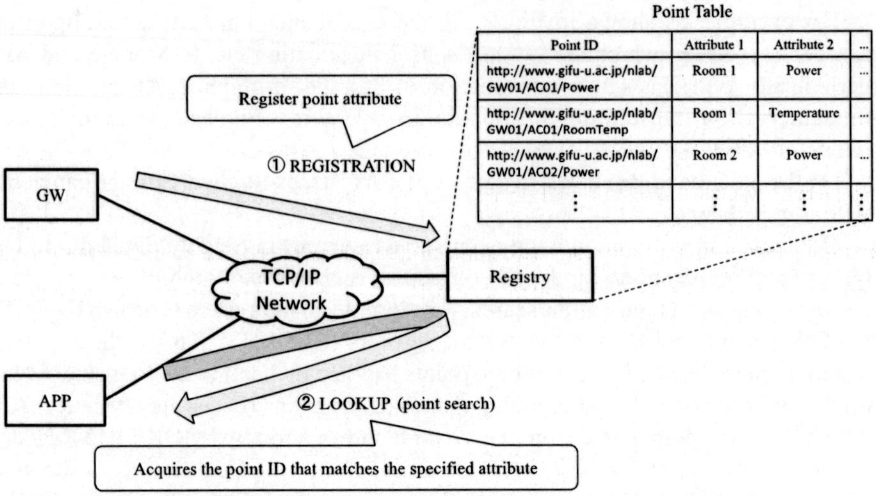

Fig. 8.23 Communication between component and registry (Point registration/search)

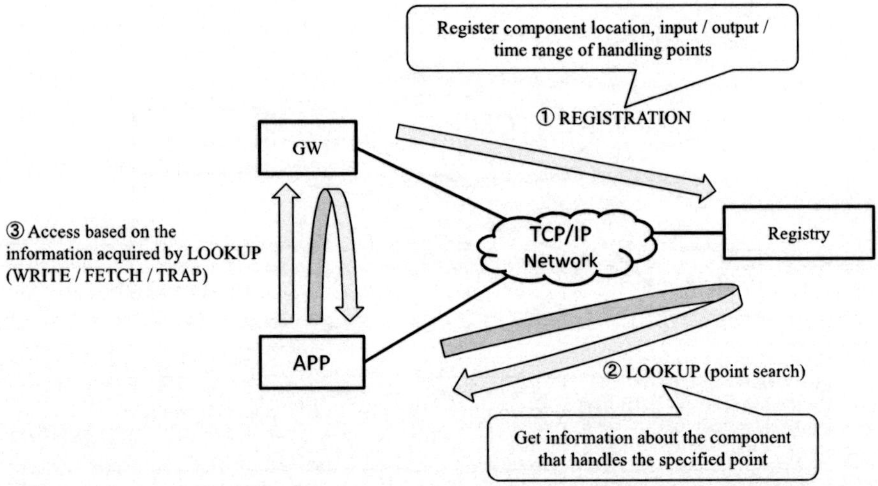

Fig. 8.24 Communication between component and registry (Componet registration/search)

supported operations (WRITE/FETCH/TRAP), handling point names, input/output Register the time range etc. in the Registry (REGISTRATION). When APP specifies a point name in Registry and LOOKUPs, APP can obtain various information of the component that handles the point.

8.3.2 Implementation of IEEE 1888 Standard Software

In the IEEE1888 implementation, the key is that you can communicate with SOAP, so you can choose any hardware or operating system. In addition, languages that are often used in web development, such as Java, C#, and PHP, can be used as development languages, libraries that use SOAP are also available, and there are plenty of references, so it will be suitable for development.. For the C language, the IEEE1888 protocol stack (communication software component) has been released by GUTP and can be applied to embedded development.

IEEE1888SDK is provided as a virtual machine in ova (Open Visualization format Archive) format, so a separate virtual environment such as VMWare Workstation Player is installed and used. VMWare Workstation Player is free for personal use. Note that IEEE1888SDK does not include a protocol stack, so it is not necessary to prepare it when implementing APP or GW.

In addition, the implementation of FIAP Storage, which was also developed by GUTP, can be used free of charge, and it is installed by default in the above SDK, but it is also provided as a single unit, and it is also possible to build an environment on your own. However, because the installation procedure is somewhat complicated, it is recommended to use IEEE1888SDK Storage when developing APP and GW.

There is no publicly available implementation of Registry as of October 2018. It should be remembered that the software provided by GUTP is a BSD license, so care must be taken when handling copyright notices for commercial use.

In the following, we will introduce the communication contents (payload) and sample program for implementing communication between the building multi-type air-conditioner and the aggregator in IEEE1888 assuming the construction of a virtual power plant with the building multi-type air-conditioner. Figure 8.25 shows the assumed system configuration. In Fig. 8.25, the building multi-type air-conditioner is composed of a set of outdoor unit AC_{nn} and indoor unit IU_{mm} (*nn* and *mm* are arbitrary identification numbers). In the building multi-type air-conditioner, the indoor unit and the outdoor unit communicate with each other using a communication protocol unique to the manufacturer. IEEE1888 GW is installed in each building and plays a role of converting air-conditioning data in the building into IEEE1888 protocol.

The aggregator functions as Storage of IEEE1888, and GW WRITE/FETCH the points shown in Table 8.13 to the aggregator. The point name has the following structure. http://main organization name/sub organization name/aggregator name/gateway name/air-conditioner name/indoor unit name/value.

In Fig. 8.25, GW writes the power consumption of the building multi-type air-conditioner and the measured room temperature of each indoor unit to the aggregator (Storage), and FETCHs the power limit command value.

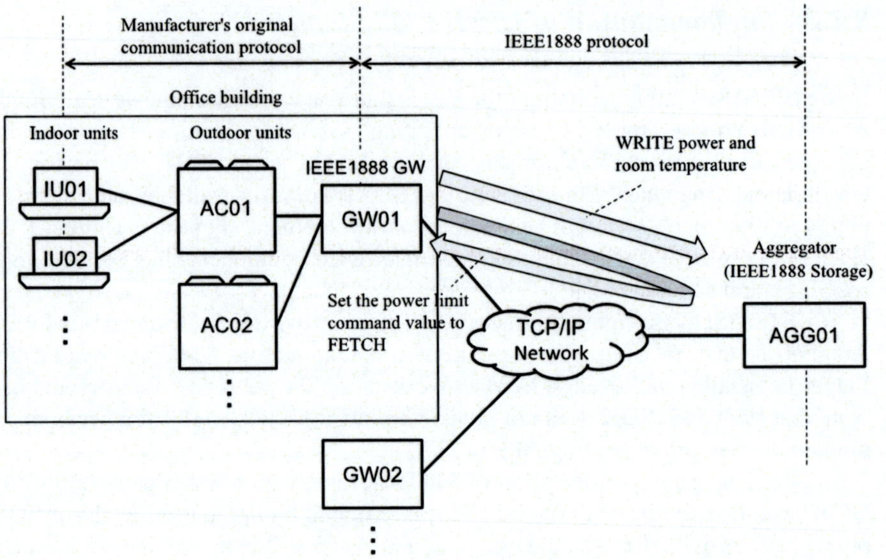

Fig. 8.25 System configuration of buildein multi-type air-conditioning virtual power plant using IEEE1888 communication

Table 8.13 Point table of building multi-type air-conditioner

Point name	Meaning	GW operation
http://www.gifu-u.ac.jp/nlab/AGG01/GW01/AC01/P	Power	WRITE
http://www.gifu-u.ac.jp/nlab/AGG01/GW01/AC01/IU01/Ta	Temperature	WRITE
http://www.gifu-u.ac.jp/nlab/AGG01/GW01/AC01/IU01/L	Power Limit command	FETCH
…	…	…

An example of WRITE payload is shown in Fig. 8.26. The payload is roughly divided into three parts: HTTP request line, header, and message body. The message body is described in XML format, and the s: Envelope tag to s: Body tag are descriptions of SOAP communication.

An example of WRITE payload is shown in Fig. 8.26. The payload is roughly divided into three parts: HTTP request line, header, and message body.

The message body is described in XML format.

From the s: Envelope tag to the s: Body tag is a description of SOAP communication.

Furthermore, data related to IEEE1888 communication is stored inside the dataRQ tag.

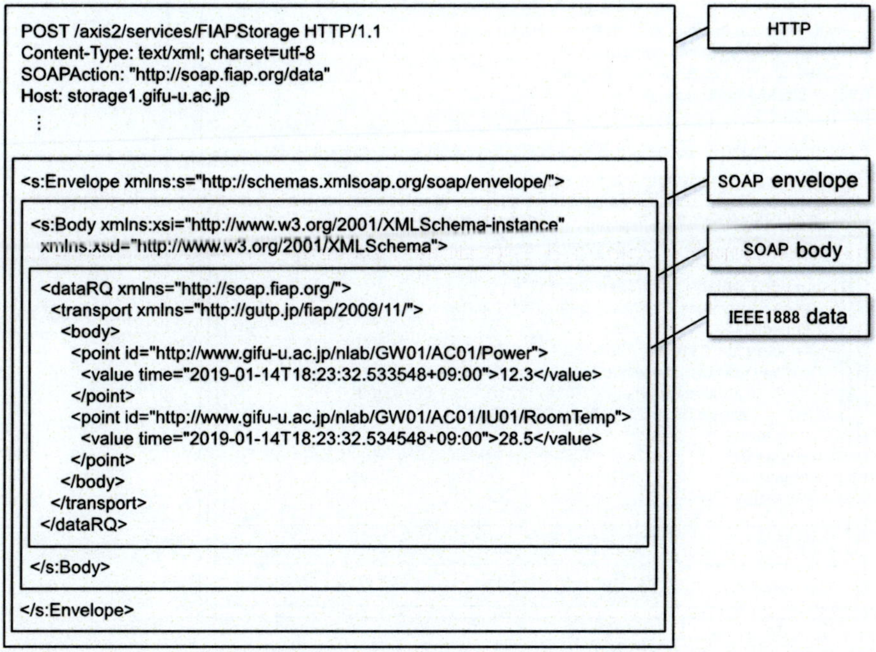

Fig. 8.26 IEEE1888 WRITE payload example

dataRQ specifies a call to the data method. Data related to processing requests and processing results are stored inside the transport tag. The transport consists of a header and a body. The header stores the point search conditions for FETCH and the success or failure of the process. The body stores the data to be processed or the response data.

In the case of WRITE, there is no header, and the information of the point to write to is stored in the body. The point name is stored in the id attribute of the point tag, and the time in ISO8601 format is stored in the time attribute of the value tag. The value enclosed in the value tag is the point value. In this example, the power (http://www.gifu-u.ac.jp/nlab/GW01/AC01/Power) is 12.3 [kW] and the room temperature (http://www.gifu-u.ac.jp/nlab/GW01/AC01/IU01/RoomTemp) stores 28.5 [°C].

Next, Fig. 8.27 shows an example of FETCH payload. The basic structure is the same as the WRITE message. Data related to IEEE1888 is stored inside queryRQ in the message body. Unlike WRITE, there is a header tag that contains a query tag. Query is the basic unit of search processing, and Storage executes search processing for each query. In the id attribute of query, GUID (Globally Unique Identifier) is described, and Storage uniquely identifies the point search processing by this value.

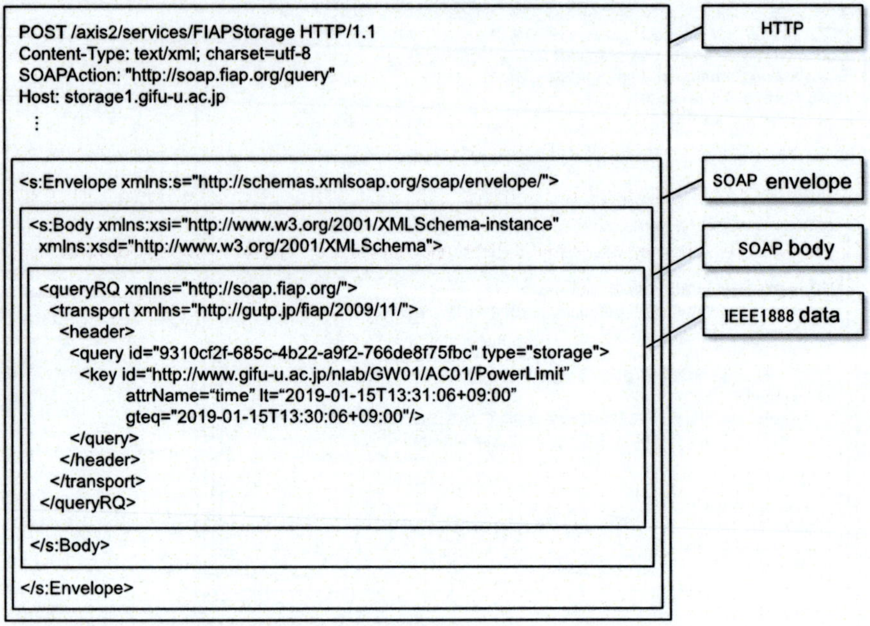

Fig. 8.27 IEEE1888 FETCH payload example

There is a key tag in query that holds search conditions. Specify the search target point name in the id attribute of key and the search target item (time or value) in the attrName attribute. If time is specified, the search is performed by specifying the time range, and if value is specified, the search is performed using the point value. Next, specify the search condition that specifies the time and the magnitude of the value. The contents differ depending on the conditions, and gt (greater than: greater than), gteq (greater than or equal to: or more), lt (less than: less than), lteq (less than or equal to: or less) can be specified.

Next, Fig. 8.28 shows an example of a response returned from Storage in response to a FETCH request from GW. The FETCH response is stored after ns: queryRS in the message body. In the header in the transport, "OK" indicating normal termination and the contents of the query sent to the Server are described. The information (point name, time, value) of the point corresponding to query is stored in body.

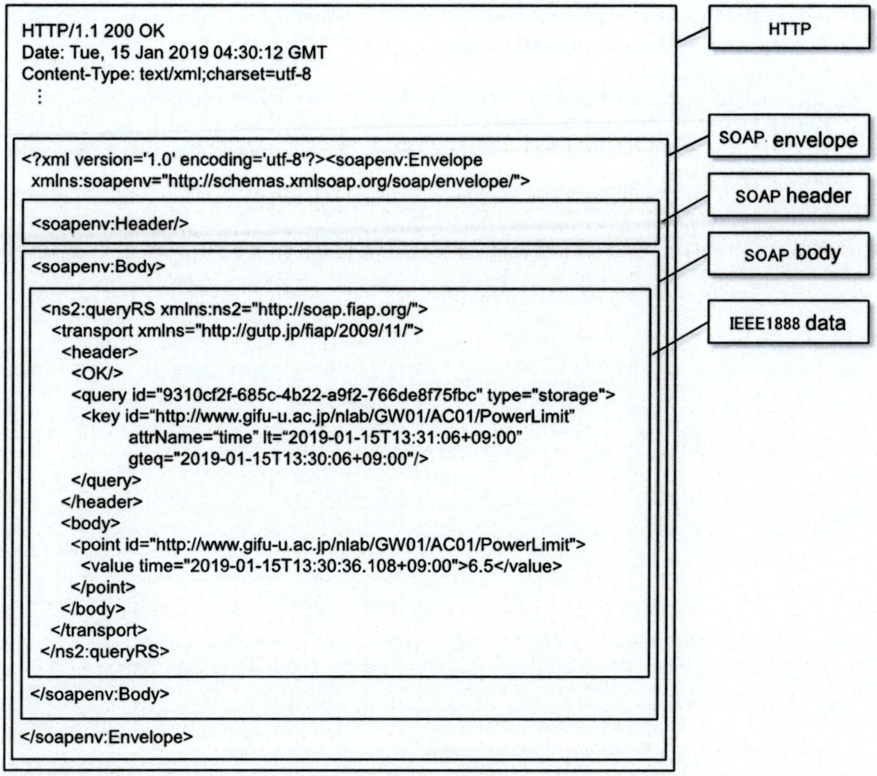

Fig. 8.28 IEEE1888 FETCH response payload example

References

1. W3C standard, *XML Schema Definition Language (XSD) 1.1 Part 1: Structures. W3C Proposed Recommendation* (2012). http://www.w3.org/TR/2012/PR-xmlschema11-1-20120119/
2. P. Saint-Andre, *Extensible Messaging and Presence Protocol (XMPP): Core. RFC61020* (2011)
3. OASIS Standard, *Energy Interoperation Version 1.0. OASIS Committee Specification Draft 01* (2010). http://docs.oasis-open.org/energyinterop/ei/v1.0/os/energyinterop-v1.0-os.html
4. EnerNOC OpenADR 2.0 VTN open source implementation. http://github.com/EnerNOC/oadr2-vtn-new, http://www.openadr.org/general-resources
5. Electric Power Research Institute (EPRI), *OpenADR 2.0b Open Source Virtual Top Node (OADR2.0b VTN) Version 0.9.7.0. Product Id3002007431* (2017). http://www.epri.com/#/pages/product/3002007431/?lang=en-US
6. IEC standard, *Communication Networks and Systems in Substation—Part 6: Configuration Description Language for Communication in Electrical Substations Related to IEDs. IEC 61850-6* (2009)
7. IEC standard, *Communication Networks and Systems in Substation—Part 7-2: Basic Communication Structure for Substation and Feeder Equipment—Abstract Communication Service Interface (ACSI). IEC 61850-7-2* (2003)

8. IEC standard, *Communication Networks and Systems in Substation—Part 8-1: Specific Communication Service Mapping (SCSM)—Mappings to MMS (ISO 9506-1 and ISO 9506-2). IEC 61850-8-1* (2004)
9. ISO standard, *Industrial Automation Systems—Manufacturing Message Specification—Part 1: Service Definition. ISO9506-1* (2000)
10. D. Steedman, *Abstract Syntax Notation One, ASN.1, The Tutorial and Reference* (The Camelot Press, 1990)

Chapter 9
Outlook of the Virtual Power Plant: A Japan Perspective

9.1 Potential Resources

9.1.1 Potential of Building Multi-type Air-Conditioning System

Taking Japan as an example, the number of installed building multi-type air-conditioners in Japan estimated to be 1.52 million outdoor units. Here, the building multi-type air-conditioner is not an entire commercial air-conditioner, but a part of a packaged air-conditioner in which multiple indoor units are piped to an outdoor unit. There are about 5–10 indoor units per outdoor unit of a building multi-type air-conditioner, the order of indoor units is estimated to be 10 million units. There are various types of outdoor units with different air-conditioning capacities, but 10 H.P class, which is the representative model, that is, the rated cooling capacity is about 28 kW, the rated power consumption per unit is about 10 kW. It can be said that the total installed power is about 15 million kW. Of course, since it is partly operated on a daily basis and the average power consumption is less than the rated value, it goes without saying that only a few percent can be used as negative wattage resources. However, there is no doubt that the installed capacity is a useful load type.

Let's estimate the number of customer buildings where these multi-type air-conditioners are installed. Building multi-type air-conditioners are mainly used for medium and small business buildings such as office buildings, commercial facilities, schools and hospitals. One of the official statistics on the number of non-residential buildings and total floor area in Japan is the Basic Survey on Land and Buildings of the Ministry of Land, Infrastructure, Transport and Tourism. This is a large-scale statistical survey conducted every five years, and the latest version is published in 2013.

C. Ninagawa, *Virtual Power Plant System Integration Technology*, Power Systems,
https://doi.org/10.1007/978-981-16-6148-8_9

Here, it is assumed that the buildings are small-to-medium-sized building where the building multi-type air-conditioner is installed, and a medium-sized building that can be a target of the FastADR virtual power plant system. When applied to the classification of the same statistics, buildings with a total floor area of 2000–10,000 m^2 are considered to be applicable. The total number of medium-sized buildings nationwide is 140,000, the total floor area is 580 million square meters. In this way, it can be seen that the building for which the building multi-type air-conditioning system is installed.

Of course, among these buildings, only a small percentage of the total customers can be subject to FastADR, that is, they can temporarily suppress power. In addition, these are the total number of nationwide, and it should be considered only in urban areas where some accumulation is seen. Therefore, it is reasonable to consider that the power that can be suppressed as a virtual power plant, that is, the negative wattage generated power is less than half of that.

Based on the above assumptions, it seems that, even if it is limited to office buildings, the installed capacity of the building multi-type air-conditioning FastADR is the potential of several large-scale gas turbine thermal power generators of the electric power company. It is not difficult to imagine that if we gather building multi-type air-conditioning customers other than office buildings, such as factories, schools, shopping centers, and public offices, it will become a full-scale virtual power generation resource.

9.1.2 Suitability of Air-Conditioning Load for High-Speed Demand Response

By reducing the time granularity, it is possible to manage the consequence of room temperature change, precisely control the amount of power suppression, and reduce the uncertainty of response. Compared to the control certainty of traditional large-scale thermal power plants and hydroelectric power plants of huge dams, that is, the certainty of response to control commands from the central power supply command station, the power suppression response of individual air-conditioners is uncertain.

The author's basic idea is that a certain amount of power suppression can be ensured as an averaging effect by suppressing hundreds, thousands, and tens of thousands of air-conditioners for a short period of time to reduce power consumption. Of course, the repeatability of a centralized unit is not 100%. According to a study by Lawrence Berkeley National Laboratory in the United States, the negawatt power generation with a large number of air-conditioners has been reported to be inferior to the response success rate of a centralized large-scale power plant [1–3]. Furthermore, when large-scale natural disasters such as earthquakes and tsunamis occur, the combination of centralized and distributed types may avoid long-term catastrophic situations.

There is a discussion, such as, the air-conditioning load of about 1 h is predicted based on the temperature of the next day, etc., and if the heat load is small, a large amount of air-conditioning demand response can be expected. In the case of equipment such as an air-conditioner with a heat storage layer, the argument would be that the air-conditioning power consumption can be reduced by stopping other air-conditioners if the heat load is small while cooling from the heat storage layer.

However, it is unlikely that such an expensive heat storage device is equipped in the building multi-type air-conditioning system of general buildings that is widely used as envisioned in this book. When there is no heat storage device, the amount of air-conditioning power consumption in the baseline is small in the weather condition time zone where the heat load is small. So, it seems that there is often a small amount of suppression from the point where power is not originally used. In cases where the weather conditions of the next day are expected to be severe, the air-conditioning power consumption will be higher than usual for several hours during that time period. On the other hand, the amount of power that can be suppressed is increased under these conditions. So, it is desirable to demonstrate air-conditioning capability. This trade-off will be optimally adjusted in the context of economic incentives.

As described in this book, as a virtual power plant, in order to ensure controllability equivalent to that of a power plant from a group of building multi-type air-conditioners, hundreds and thousands of ordinary small and medium-sized buildings are aimed at improving the certainty by the averaging effect. Building multi-type air-conditioners installed in such customers are very general standard mass-produced products and do not come with a heat storage tank. Therefore, demand response is not suitable for such a general building multi-type air-conditioning system with a slow that suppresses air-conditioning power for many hours.

The FastADR can be terminated and released quickly in 5 or 10 min, depending on the condition of the air-conditioning section in which each building multi-type air-conditioner is responsible, and then rotated to another building multi air-conditioner. In such a system, the fast response to air-conditioning demand response is not the demand response by time-based human activation as in the past, but the FastADR converted to M2M in minutes.

9.2 System Design for Practical Application

9.2.1 System Design for Supply and Demand Adjustment Market: Japan Case

By drawing from the experiences of the Japan, the author illustrates the Japanese regulatory approach, while improving the reliable operation of the grid.

In recent years, there has been rapid progress in the separation of dispatched electricity. The Japanese electric power wholesale market JPEX introduced the next day spot market and the market before the day. These markets also have been booming.

By the way, at the February 2019 stage when this book was written, the total daily power consumption of Japan as a whole was about 2.5 billion to 3 billion kWh, about one quarter of 500 million to 800 million kWh was JEPX wholesale power.

In addition to the wholesale market, a yearly procurement open call system is in operation for each power system area, but this is a wide area that crosses the area due to the separation of dispatched power, so the supply and demand adjustment power [ΔkW] and power capacity in real time [kW] and electric energy [kWh) supply and demand adjustment market system is under development.

The supply and demand adjustment market was initially aimed at opening in fiscal 2020, but at present, wide-area procurement and operation are expected to be carried out for each product sequentially from fiscal 2021.

The virtual power plant system assumed in this book is based on the premise of the integration of a large number of building multi-type air-conditioners and FastADR of power storage batteries. It is a high-speed, automated demand response that responds in a matter of minutes with the fast response of the storage battery helping the response characteristics of the building multi-type air-conditioning system. In the previous section, we described that a system that rotates a large number of short-time demand response durations is suitable for demand response of building multi-type air-conditioners. The consequences of demand response over many hours is the disturbance of comfort zone, and power cannot be continuously suppressed for hours after rotation.

9.3 Key Social System

9.3.1 Compensation for Virtual Power Plant Service

Even if innovation technology to realize a virtual power plant is in place, it cannot be implemented in society unless a system for economic trading of its value is in place. In particular, in order to buy and sell "virtual power products" that were not in the conventional concept of "negawatt". Among other things, how to deal with the measurement method that is particularly deeply related to power trading in "negawatt trading" will be a big problem.

This book has described a virtual power plant based on the aggregation of building multi-type air-conditioners and power storage batteries. The feature is to distinguish the power consumption of the building multi-type air-conditioner group and the charge/discharge power of the power storage battery group from the power consumption of other consumer loads from among the large number of consumer groups spreading in the region, and specify the amount of negative wattage. Therefore, a system that settles the economic value as compensation is required.

Electric power companies set up transaction meters that have been certified as verification meters for each customer and perform meter readings. This will be a smart meter in recent years, but it is still a measurement settlement for each customer. At present, even a smart meter is a settlement unit with a long time granularity of 30 min of energy. In the future, the time unit may be shortened to 15 min or several minutes. In addition, not only the amount of power [kWh] for each time frame such as 30 min, but also power [kW] may be measured at shorter time intervals. However, in any case, as long as there is one transaction meter for the entire consumer under the current legal framework, the air-conditioning load power and the power storage battery charge and discharge are not included in the consumer power or power consumption.

To put it simply, even if building multi-type air-conditioning reduces air-conditioning power consumption with FastADR in an office building, which is a single electricity bill contract unit. For example, FastADR power suppression cannot be measured with one power transaction meter for one building. As described above, the aggregator requesting the virtual power plant FastADR is based on the premise that the incentive will be paid to the consumer as a price for generating negative wattage due to power suppression. Therefore, it is necessary to separate and measure only the power of the equipment group that the aggregator and the customer have identified as FastADR, that is, the building multi-type air-conditioning equipment and the power battery equipment in this book, and settle it between the aggregator and the customer.

One solution is to install a separate power meter for only the FastADR target facility power, measure it, and transmit the data from the customer to the aggregator to separate and identify the negative wattage. However, originally demand response has to estimate the amount of suppression against the baseline estimate. As clarified in Chap. 4 of this book, accurately estimating the baseline itself is very difficult in advanced data science. At present, the baseline is set as an alternative value that determines the calculation method as a system such as the average of the past. Therefore, even if the power at the time of demand response is accurately measured with a wattmeter that has been verified and guaranteed accuracy, the baseline estimation itself is not accurate.

In addition, in a super-distributed resource group such as FastADR in a building multi-type air-conditioner group, installing a power meter in individual equipment increases costs, and depending on the amount of initial investment, water may be used to promote FastADR. In addition, the building multi-type air-conditioner is always equipped with an inverter for the refrigerant compressor, and the built-in microcomputer controller has a function to constantly measure the effective value of current consumption or power consumption in milliseconds for overload protection.

Using this function, ERC (Energy Resource Controller) or BEMS (Building Energy Management System) communicates power consumption to a resource aggregator at intervals of several minutes or several tens of seconds, and communication data according to IEEE1888 and IEC 61850. Needless to say, the power storage battery system itself can always accurately and accurately measure the charge and discharge power.

As shown in Fig. 9.1, a system that aggregates the generated negative wattage $\sum W_B$ of the virtual power plant that aggregates the building multi-type air-conditioner group and the power storage battery group, and distributes the incentive $\sum Y_B$ to each consumer is desirable. Power retailers and electricity charges will be

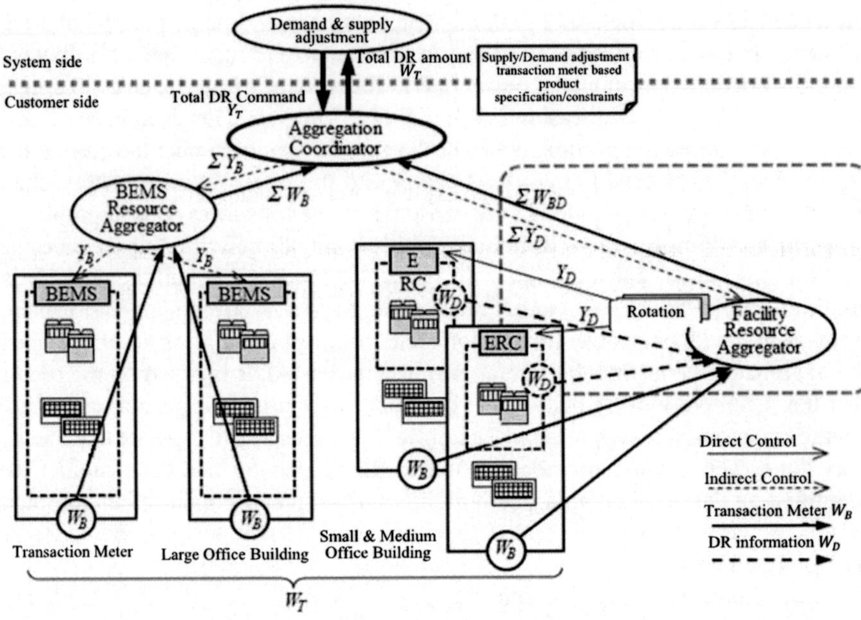

Fig. 9.1 Conceptual diagram of negative watt metering and incentive allocation of a virtual power plant that aggregates building multi-type air-conditioning units and power storage battery groups

based on one transaction smart meter per customer. However, the amount of FastADR negative wattage generated by the virtual power plant system can be obtained from the above-mentioned data from the embedded microcomputer controller. It seems desirable to distribute the incentive Y_T of the entire virtual power plant to this W_B or from W_{BD}.

In this way, in one customer building, even if the building multi-type air-conditioning power and power are suppressed by FastADR, the power consumption has increased in other equipment. So, the meter for consumer transaction smart meter W_B You can avoid the phenomenon that the value increases and negawatt cannot be measured. By transmitting only the power consumption W_D of the FastADR target equipment itself, it is possible to measure negative watts and to contribute incentive Y_D to each consumer by contribution.

Needless to say, the measurement law barrier needs to be solved, but in power system reform that is no exaggeration to say once every 100 years, it is a measurement that is suitable for ICT without hindering future innovation. It is essential to develop a settlement method.

9.3.2 Clean Value of Virtual Power Plant

Solar power generation, wind power generation, and hydropower generation are said to be safe power generation methods that are friendly to the global environment and do not have to worry about radioactivity, such as nuclear power, because there is no CO_2 emissions during power generation. Certainly, solar power generation, wind power generation, and hydroelectric power generation do not emit CO_2 and worry about radioactivity when the facilities are completed and operated. These natural energy power generations are environmentally friendly compared to power generation that burns fossil fuels such as CO coal, oil and natural gas during operation.

However, if these natural energy power generations are also strictly observed, it may be necessary to consider the impact on the land and natural environment when constructing a hydroelectric dam, as well as the environment and disasters in the area where solar power generation and wind power generation are installed.. Furthermore, even though it is naturally friendly during operation, it can be said that it is necessary to consider the earth resources when manufacturing new power generation equipment and the waste when abandoning equipment due to aging.

In this regard, the virtual power plant based on FastADR, which is the theme of this book, can be said to fundamentally reduce energy consumption because it reduces power consumption while controlling power consumption from the baseline. In other words, it is a power generation system that does not use earth resources, emits CO2, and does not emit any waste as well as radioactivity. Of course, as mentioned in this book, there are many caveats. FastADR is temporary for a short time even when it is rotated, and power consumption rebounds after FastADR ends.

In addition, since the power storage battery system, which is a component of the virtual power plant, is a physical facility, it uses the earth's resources like solar and wind power generators, so it can put an impact on the environment when it is discarded. Power storage battery systems can control power charging and discharging instantaneously and reliably, so they tend to rely on controlling the positive/negative power output of virtual power plants. However, since FastADR is a completely clean and safe supply and demand adjustment power generation, it is necessary to overcome technical difficulties so that FastADR can be used as much as possible. It should be noted that FastADR negative wattage and power storage battery positive watt are properly shared within the virtual power plant.

Needless to say, the power generation of "negawatt" in the balance of supply and demand temporarily produces effects, and it goes without saying that it is not sufficient for the base load. Base load is the amount of electric power that continuously generates energy, the so-called kWh value. On the other hand, in the supply and demand adjustment of the generated power and the demand power of the power system, it means that the instantaneous power is balanced, and it is the value to change the power, that is, ΔkW value. It is necessary to distinguish between these two values.

There is no world-wide supply and demand adjustment as much as the "negawatt" produced by virtual power plants through centralized control of FastADR. Author hope's that this technology will play an active role in the future.

References

1. J.H. Eto, J.N. Hoffman, E. Parker, C. Bemier, P. Young, D. Sheehan, J. Kueck, B. Kirby: The demand response spinning reserve demonstration—measuring the speed and magnitude of aggregated demand response. IEEE Computer Soc. 2012–2019 (2012)
2. J. Eto, C. Goldman, G. Heffner, B. Kirby, J. Kueck, M. Kintner-Meyer, T. Mount, W. Schultze, R. Thomas: *Innovative Developments in Load as a Reliability Resource* (IEEE, 2002), pp. 1002–1004
3. J. Bode, M. Sullivan, J.H. Eto, *Measuring Short-term Air-conditioner Demand Reductions, for Operations and Settlement* (2002)